THE NEW RESEARCH FRONTIERS OF COMMUNICATIONS POLICY

THE NEW RESEARCH FRONTIERS OF COMMUNICATIONS POLICY

EDITED BY

DONALD M. LAMBERTON

URBAN RESEARCH PROGRAM
RESEARCH SCHOOL OF SOCIAL SCIENCES
AUSTRALIAN NATIONAL UNIVERSITY
CANBERRA, AUSTRALIA

N·H

1997

ELSEVIER

AMSTERDAM – LAUSANNE – NEW YORK – OXFORD – SHANNON – TOKYO

ELSEVIER SCIENCE B.V.
Sara Burgerhartstraat 25
P.O. Box 211, 1000 AE Amsterdam, The Netherlands

ISBN: 0 444 82251 8

This book is printed on acid-free paper.

Printed in The Netherlands.

Preface

In no other field has the lag of policy development behind technological and market developments been as dramatic as in communications services. The lag is in large measure the result of the little knowledge there is about the economic and social significance of the new technology and its dynamics. Research can assist in reducing our ignorance with potentially immense benefits for the information society, and an assessment of research priorities is now urgently needed.

It is not often that one has the advantage of the presence of so many distinguished scholars at the one time to make possible a two-day workshop on research needs. Such an opportunity presented itself following the Tenth Biennial Conference of the International Telecommunications Society in Sydney in July 1994. A number of international delegates were brought together with local researchers and communications policy advisers in the conducive environment of University House at The Australian National University, to discuss research priorities.

A notable feature of the workshop was the fact that no papers were presented or other background documentation circulated to participants beforehand. It was organized on the basis that the participants were all active in policy advice relevant research covering the main economic and social aspects of communications policy. Some of the participants were invited to lead the discussion on particular topics and the rest was left to group dynamics.

The result was an intellectually exciting and highly productive two days of discussion of emerging communications issues, the level of ignorance of both analysts and policy-makers and the areas of urgently needed research.

The papers in the volume do not reflect adequately the full richness of the insights and ideas the workshop generated but they do provide a valuable starting point for policy-makers and researchers who are grappling with the complex economic, social and policy implications of rapid developments in communications technology and in communications markets.

Don Lamberton is to be commended for his initiative in organizing the workshop and the sponsoring organisations for their support and encouragement.

Maurice Haddad
Bureau of Transport and Communication Economics

Canberra, July 1996

Contents

CHAPTER 1

Research frontiers

Donald M. Lamberton

Introduction

After the Tenth Biennial International Conference of the International Telecommunications Society (ITS) in Sydney in 1994, we were able to organize a two-day workshop at University House, Australian National University, to address research priorities for telecommunications and information activities. The intention was to take advantage of the presence in Australia of distinguished scholars to hold discussions of the research that needs to be done into the economic, social and policy aspects of these elements of the information society. A list of participants is provided at the end of this volume.

The ITS is an association of professionals (academics and practitioners in operating companies, consultancies, government agencies, international organizations) with an interest in the growing field of telecommunications planning, policy formulation and economic decision analysis. As a world-wide network, it seeks to foster research, dissemination of research findings and policy debate. The Tenth Biennial ITS Conference has hosted by Telstra Australia with support from Alcatel Australia, AT&T, BT, the Australian Department of Communications and the Arts, Booz-Allen & Hamilton, NSW Office of Economic Development, NTT, Siemens Australia and University of Technology, Sydney.

The Workshop was sponsored by the Bureau of Transport and Communications Economics; Urban Research Program, Research School of Social Sciences, Australian National University; Faculty of Information Sciences and Engineering, University of Canberra; and the National Library of Australia.

Workshop topics

There were no papers presented at the Workshop. Seven major topics had been identified and chairpersons, selected for special interest, opened the discussion:

Demand Studies (Don Lamberton)
Networking (Cristiano Antonelli)

Technological Change (Peter Swann)
Market Structure (Mark Schankerman)
Growth and Development (Rod Maddock)
Regulation (Martin Cave)
Social Dimensions (David Allen).

In all but two cases the corresponding chapter in this book was subsequently written by the chairperson.

The general title of "Telecommunications and Information Activities" was adopted quite deliberately in order to put telecom economics into a wider setting. This was not wholly successful as some participants took the view that social aspects were being isolated and marginalized — and I shall return to this problem later in this chapter.

The research process

A recent issue of the Northern Business Information *Datapro* catalog of research reports offers 15 reports at a total cost of more than $70,000. It ranges widely: for example, the US Voice Processing Market; the Chinese Telecommunications Market; European, Asian and Latin American Cellular Operators; Broadband Markets. The sales pitch claims these reports provide both the facts and the real-time perspective and insight that make them useful. But it is my impression that the *Datapro* reports are rather like the great bulk of the communications magazines and newsletters in that, of necessity, they are preoccupied with fact rather than perspective and insight. The justification is that so much is happening, with decisions coming in rapid-fire order, that it is as much as we can do to keep up with events.

Psychologist George Miller described the earlier phases of this development in these words:

> Communication has become an ... obsession. Those who admire it justify the obsession as a necessary precondition for democratic government, or for a free economy; others view it less favourably as a defence against the loneliness of personal independence, or against the insecurity of competitive co-existence. Whatever the reason, the vast communication industries that have sprung up in this century now bathe every willing citizen in a continuous and unprecedented flow of written, spoken, musical, and pictorial messages (Miller, 1983:ix).

In such circumstances, information is clearly at a premium. Predictions command a high price; and the deeper understanding of the complex changes that might be generated through research tends to fail to keep up with the hectic pace. Is this as much as we should aspire to? Sune Carlson prefaced his essay, *Swedish Industry Goes Abroad* with the following words: "If this essay enables the reader to see a little better where and how the Swedish export industry has been until now, my ambition will have been fulfilled" (1979:6). But we seek from history some guidance for future decisions and so we have the research industry.

The research output overseas and in this country is scattered. There is no readily available compendium that brings together the research findings from, for example, WIK in Germany, IUI in Sweden, IDEI in France, CIT-I at Columbia University, SPRU in the UK, the East-West Center in Honolulu and BTCE in Australia; nor is it a simple matter to find out what research is in progress. Where are the directories of current research? Where does one find the lists of current doctorate theses?

I suspect there are several propositions that would command wide support — although not necessarily from the same groups. First, there is too little research being done. Second, the results of research are not disseminated widely enough. Third, too little use is made of research findings. But before we try to change this situation, we must take care in shaping our idea of what constitutes research.

As I have implied, it is important to distinguish between the news material and research. Our difficulty in keeping up with the news should lead us to seek better basic data. This too is not easy and it is costly. Statistical agencies have always found it difficult to initiate coverage of new developments and in the communications field there are added obstacles: the intangible nature of information and consequent troubles with output measurement; the complexity of property rights; and the increasingly proprietary nature of the data. We must envy those involved in the US National Telecommunications Demand Study with its data set of more than two million observations on both business and residential customers (Taylor, 1994). I should add that the NTDS study is a joint undertaking of two data/consultancy firms (INDETEC Inc. CA and PNR, PA) with Lester Taylor of the University of Arizona. I suggest it is a prime responsibility of all those who claim to be advancing research in this field to pave the way with surveys and generally support data collection. Without data, there is no empirical research, and we are left in a situation, which some seem to relish because it permits endless debate and rehashing of old arguments.

The analysis part of the research process must leave room for both theory and empirical investigation. The key requirement is simply that the groups involved keep in touch. Funding bodies should try to bring this about through their progress assessments. It is my view that the organization of research needs to permit the early identification of worthwhile work and then to facilitate both communication within the research community and communication to the much wider groups that are, or should be, interested.

All too slowly there seems to be recognition of the complexity of the process of innovation in industry. The process of implementation of communications research findings by industry and policy-makers shares that complexity. Just as the data collection can be a slow and costly process, so can research implementation. It is not achieved by a few seminars and conferences, nor can success be measured by the presence of some CEOs. Learning is hard, especially when it involves both giving up old entrenched ideas and resisting significant "technology push". We need research, surveys of research, translations of research. Those responsible must remain true to the research ideal; their efforts must not degenerate into the selling of technology — and I confess that is my reaction to many recent books, articles and TV programs — or, worse still, the selling of their own enterprises — be they consultancies, "research institutes" or "universities". And we need interaction, real interaction between the various participants.

This is not to argue there is no room for specialization. It is the essence of economic development but there are limits. Neither analysts nor policy-makers can afford to be out of touch with reality; empirical researchers need to know their data. A very good practitioner of applied economics said:

> any intelligible account of transactions, any reasoned history, implies a theory: any user of our studies is bound to be concerned with the firmness and representativeness of the evidence, and the rigour and generality of our reasoning. It is here that the impulse towards realism may be defective. When we recreate a scene before our eyes, we are not abstracting, we are not reasoning: we may be missing connections and constraints, we may make statements that analysis would show to be downright fallacies. But for the type of mind that likes to see things as they are, that sort of hard thinking may be uncongenial; and the realist's criticism of the unreality of theory may only provide a rationalisation of his own reluctance to make the mental effort that theory would require of him (Phelps Brown, 1980:11).

The postmodernist critique suggests unstable meanings and unknowable truths but it has not relieved researchers of these obligations.

I have tried to indicate some of the steps involved in the research process, the spirit in which it needs to be conducted, and some of the conditions that might ensure use of its findings. While I have set some constraints on what qualifies as research, I still have not really said what is its core.

Back in the Henry Mayer days, the periodicals notes of *Media Information Australia* (May, 1990:112) said there were more than sufficient journals and newsletters on telecommunications alone, but that one Australian journal, *Prometheus*, had the distinction of making and publicizing interconnections — and that would be the sub-title I should like to give this chapter.

Research is the attempt to find interconnections: between actions and events; between technologies; between telecommunications and other information activities; between groups of people; between the economic and the non-economic; between different kinds of information; between policy interventions and consequences. And we should set quality standards. We are not always (perhaps only sometimes) satisfied with interconnections that meet the econometrician's tests of causality. We hanker after more than the predictability of statistical "laws" (Geweke, 1984:1102-1108).

Interdisciplinary messages

At BTCE Forums and the post-ITS'94 Workshop from which this book has evolved, and I am sure at countless other research gatherings around the world, a division between what are labelled erroneously the economists and the others, was apparent and is a matter for concern. These labels are not good enough. The Econ Tribe shows a great deal of diversity: micro and macro theorists, economic historians, econometricians, with many specializations — industrial organization, industrial relations, development economics, international economics, regional economics, health economics, information economics. Each defends their own territory; they resist change; and they view the world with their own particular kind of blinkers.

The same goes for the disciplinary groups that make up "the others": communication studies, information science, psychology, sociology, law, history, organization studies, management, engineering. They too wear their blinkers. The relationships between disciplines are competitive but we must assume a shared objective of unearthing interconnections, of achieving better understanding of the roles of information and communication activities in the hope of better decisions, better use of resources and larger GNP. If that seems an economist's formulation, recast it as *Gross National Happiness* (Menou, 1994).

The main message is that while truly interdisciplinary research is needed, there is room for specialization. A not-so-recent but valuable catalog of reasons for interdisciplinary effort is available (Streeten, 1974):

(i) A practical problem, e.g. a televillage project, may use input from several disciplines. They are brought together to solve a set of practical problems. The management of the project calls upon all relevant knowledge and skills but the methods of the disciplines are unchanged.

(ii) The next category involves technology transfer between disciplines. Certain assumptions, concepts or methods may be transferred. We have seen this happen between economics and political science, e.g. maximizing behaviour and the modelling of uncertainty, or between communication studies and development economics, e.g. research on technology transfer. It seems to happen less frequently in the cases of psychology (and anthropology and sociology) and economics — so infrequently that one commentator (Akerlof, 1984:2) has likened the situation to the reluctance of French cooking to use seaweed or raw fish.

(iii) A third possibility is that the justification for separate disciplines does not hold. A justification might be that the interconnections between the variables studied by, e.g. economics and communications studies, are weak and damped. If so, communications research can investigate its interconnections without taking account of those investigated by economists. Some interconnections are more important than others; it is an empirical matter.

From the research perspective, the important achievement is recognition that a wider analytical framework is necessary. When this happens, two possibilities exist. First, the disciplines can merge to create a new discipline; or, second, the old tools (concept and models) are retained but changed. The politics of research tends to favour the second possibility. For example, the economic concepts of investment and capital have been concerned with physical equipment, plant and stocks used to raise future output to a higher level. At least as far back as Adam Smith, skills were recognized as contributing to that process and in modern times we have assigned a role to human capital. The information economy discussions have brought further extensions and inclusion of information capital, organizational capital and infrastructure.

Once the appropriateness of interdisciplinary effort is acknowledged, tolerance and management are essential. Tolerance and a willingness to gain some understanding of the other relevant disciplines are necessary conditions. If it is true "that the only forum where interdisciplinary studies in depth can be conducted successfully is under one skull, and that such skulls are scarce" (Streeten, 1974:21), there are serious management implications. Creating an interdisciplinary research resource, with the necessary blend of independence and cooperation, will be a slow process. A Noah's Ark approach to recruiting and an insistence on immediate, interdisciplinary research findings is, I believe, doomed to failure. In contrast, MHRI (Mental Health Research Institute) at the University of Michigan is testimony to well-managed success.

I should like to take two illustrations of the scope and potential of interdisciplinary research. The first bears on government intervention; the second concerns telecommunications demand.

The tyranny of triangles

The majority of economists have taught, and probably at some time made use of the concept of consumer or social surplus. It seems simple enough. A downward sloping demand curve implies that all buyers except the marginal one get the good concerned at a price below what they would have been prepared to pay. So there is an area under the demand curve which sums all those differences and shows the net gain to society from the sale of the good.

Impose a tariff and the price goes up. The government has acquired part of the surplus and the surplus is reduced. But there is a little triangle lost altogether.

This reasoning has been used for a long time and much effort has gone into measuring the small triangles. So much effort as to justify *The Economist* (1994:65) using the title, "The Tyranny of Triangles". However, the revealed size has always been a bit disappointing — a recent very macro estimate came up with 1% of gross world product as the loss that could be avoided through world trade negotiations.

There has always been an opposing point of view that said these are static triangles and they might be (far) outweighed by dynamic considerations, e.g. new technology and new products. This approach has been now been revived (Romer, 1994) but given a twist that purports to show that the triangles lost are much bigger than even the most ardent free trader had ever argued. In effect, the dynamic benefits judged achievable through world "best practice" but not achieved are added to the triangle loss.

This new analysis is undoubtedly correct to argue the importance of the potential benefit from "the adoption of new technologies, the provision of new types of services, the exploitation of new productive activities, and ... imports of new types of capital goods and produced imports" (Romer, 25). However, the suggestion is that "a rough guide to the welfare losses in any country will therefore be the difference between the range of productive inputs that are available there and the range of productive inputs that could be put to use there" (Romer, 30).

A first reaction to this analysis can be one of pleasure that dynamic considerations are being given a major role in the battle against the tyranny of the old-fashioned triangles. However, we must part company when this gap between actual and potential output is described as the "true costs of badly designed government interventions" (Romer, 25). Given the complexity of the processes of innovation and technology transfer — and the absence of a lobby favouring free movement of labour — it seems unwise to invoke as a single explanatory factor explanation the role of government intervention.

Note that in his discussion of the concept of consumer's surplus, Marshall endorsed Wagner's idea that this benefit derived from "the sum total of the technical, economic, social and legal conditions" (Marshall, 1925:125, n.1). The new expanded, dynamic triangles analysis tells us history matters (Romer, 37) but government interventions are not the only determining force and they may be costly and difficult to manage. There is no tap that policy-makers can turn on to ensure the new resources and products will flow to Bangladesh, Australia or any other country as and when decreed.

For that reason there remains an important role for interdisciplinary research into the role of information in relation to, e.g. policy intervention, innovation, technology transfer and trade and investment between nations. For example, when the Group of Seven announced in 1994 a telecommunications conference with the objective of deploying new technology to spur economic growth, some officials argued that in the future business would invest only in countries that connected to information superhighways (*Wall Street Journal*, 1994). Who was in a position to test that proposition?

Telecommunications demand

My second illustration of the scope and potential of interdisciplinary research is drawn from the new version of the bible for telecommunications demand researchers, Lester Taylor's *Telecommunications Demand in Theory and Practice* (1993).

It has been a long-standing practice to emphasize the role of externalities in the growth of networks. The spreading of costs and the growth of the externalities give rise to the concept of critical mass. Until the size of network increases so that the rising utility comes into balance with the declining cost, either government or the network operator must provide a subsidy. In most treatments, the externalities are assumed to continue to add to utility as the system expands. Costs can be expected to rise as less densely populated and more remote areas are served. Noam, for example, argues that the breakdown of networks comes not by their failure but by their "very success — the spread of the service across society and the transformation of a convenience into a necessity" (1992:42).

One of Taylor's findings that is very interesting is "that call externalities may play a role larger than simply helping determine who pays for a call. In particular, there is accumulating evidence that calling is a reciprocal process not only in terms of 'taking turns', but also in terms of call stimulation. Simply put, calling appears to give rise to further calling" (p.231). The stimulation effects led Taylor to identify "an independent phenomenon", which he thought should not be treated as an effect of the externality but rather labelled the dynamics of information exchange.

He attached importance also to the distinction between calls related to the earning of income and calls that are simply a way of consuming income (p.262). This enabled him to interpret some of the elasticity of demand results and relate them to the spatial dimensions to calling patterns.

This in turn points to the need to know more about the communities of interest bound up with the physical and attitudinal characteristics of calling behaviour. What constitutes communities of interest and how are they formed?

What makes them cohere or disintegrate? How are new information goods and information services adopted? How do the stimulation effects work? Do emerging EDI patterns reflect business communities of interest? Is EDI constrained by a need to effect both standardization of messages and protocols and of coordination processes between and within firms (Brousseau, 1994)?

This list of questions can be extended. All I wish to bring out here is that questions like these require a pooling of skills, cooperation, the borrowing of concepts and methods — and perhaps even a new discipline. To some extent it exists already. There is research that is working outwards from each discipline and cooperating with others. I have pointed to the revised triangles approach to regulation and telecommunications demand contributions as illustrations but there are others. The Canadian IDRC project (Menou, 1993) goes back to scratch and tries to develop methods for the study of the impact of information in development. There are suggestions (e.g. Barlow, 1994; Lamberton, 1996) that we abandon our all-purpose concept of information as a kind of lubricating oil and develop a taxonomy of information that acknowledges its many characteristics. This would need to be paralleled by a new view of competence — especially competence in information-handling.

Such objectives are being pursued. A recent annual meeting of the American Society for Information Science in the US had a broad interdisciplinary program with the theme, *The Economics of Information*. It catered for the information industry, government and academia — for practitioners and researchers.

At about the same time the France Telecom CNET and IDEI were meeting to discuss recent theoretical and empirical advances in the analysis of network competition, with special emphasis on telecommunications. These were followed by another international congress on *The Economics of Information* organized by École Nationale Supérieure des Sciences de l'Information et des Bibliothèques.

While such meetings continue to display the separation between the economists and the others, I think there are healthy signs that tolerance and management are yielding a pay-off through communication and cooperation.

Research agenda

In the following pages, some suggestions by the Workshop participants for research priorities, grouped under the seven topics, are set out.[1]

1. Demand studies

Applying the characteristics approach to the demand for communications.

Can we improve our measurement of telephone demand by regarding it as an intermediate good, i.e. the improvement it gives to consumer happiness and business efficiency?

What are the fundamental elements of demand we can model and forecast? How can diffusion models help us forecast telecom demand? To what extent is telecom demand created, i.e. non-static utility function? Case studies are needed in each of these inquiries.

Information economics and telecommunication economics have a common problem: How to handle the dynamics of demand and supply externalities generated by complementarity, interdependence, quasi-divisibility, spillovers, and *ex ante* coordination.

Is the need for telecommunications the same as the demand? Are needs and demand met/satisfied in the same way? How can measures of demand be refined to take account of the needs of end users — both individuals and firms?

How can non-economists' social, theoretic frameworks shed light on telecom demand? The contrasts between individualistic utilitarianism and social constructions of environment may be so extreme as to prevent any contact.

There is a need to better understand demand for mobility in telephone demand. A possible structure is to establish a model to look at specifying characteristics along two lines: (i) permanent/stable characteristics such as the

1. Michel Menou contributed much of the agenda material.

need to communicate as a social activity and (ii) characteristics which may be learned and which may evolve through time. Another approach would be to explore models to transform those characteristics into telecommunications services and functions.

There is a need to revise and expand approaches to segmentation. There should be greater use of tools such as the US Bureau of the Census survey of telephone penetration which points to very uneven telephone penetration/ usage patterns.

The demand for new services has received too little attention.

What are the principal gender differences in the use of telecommunications?

What determines firm demand for telecom services? Understanding of how telecom interacts with overall firm activities is clearly underdeveloped. The scope for locating firm activities geographically is influenced but exactly how and what are the requirements on telecom for widening the locational opportunity set of different firm activities is not known.

Demand for what? Sources of change in demand? How can product innovation be modelled? Who pays for universal service? How soon can infrastructure cost be repaid? Can a universal service system be sustainable?

To what extent is demand conditioned by macro-level socio-economic trends such as urban concentration, globalization, changes in household size, etc.?

To what extent is demand induced by supply, which itself reflects technological development?

What is the nature, role and extent of imitation (fashion) in the creation of telecom demand?

2. Networking

Exploration of the dynamics of information networks:

- Communication and information exchange, together with learning, generate knowledge;
- Knowledge is a quasi-public good with spillovers and quasi-divisibilities;
- Networks are institutions that make it possible to handle the exchange of information;
- Demand and supply externalities interact in the evolution of networks.

The evolution of networks towards networks of networks can be perceived as both the outcome of technological change and regulation. Therefore, interconnection charges and standards become key issues.

In relation to standards for interconnection: (i) What institutional change is needed to ease the standards setting process? (ii) What is the right amount of standardization?

There are many questions in the political economy of networks — efficiency, access, property rights:

How does interconnection affect innovation and universal services?

What is the nature of the links between telecom networks and organizational structure and culture?

Does the introduction of telecom networks in firms enhance productivity, job satisfaction and market position?

Information-as-a-commodity can be stored, manipulated, transmitted by telecom networks. How can information-as-meaning (i.e. the half-baked ideas, the creativity of which leads to real development in a strategic sense) be "captured", stored and used in organizations? What processes, strategies, mechanisms are used to ensure that the intellectual infrastructure of the firm complements its technological infrastructure?

There is a need to incorporate the network dimension of production which is neglected in conventional economic analysis.

The characteristics of different networks (e.g. telecom vs. transportation where time plays very different roles) have to be established/evaluated.

A network of networks depends on network management. Can this be decentralized? Centralization needs to be better understood.

More modern theories of the MNC tend to treat the firm as a network — where subsidiaries represent options of future states of nature. What role does telecom play in the possible dynamic optimization of the network of subsidiaries in MNCs.

What is unique about networks?

Patterns of governance structure within the various tvpes of networks. Conditions for effective governance. Protection of citizens' rights. Equality of access to and participation in networks among countries.

3. Technological change

Measurement and classification of technological change.

Is intellectual property protection too strong?

What are the impacts on national R&D efforts of the ways in which telecom R&D is organized?

Technological change in information and communication is systemic in that it involves changes in the structure of networks and the behaviour of each agent.

What are the appropriate measures of technological change on all dimensions — social, individual, political, cost-benefit, cultural — in a firm?

How can measurement models be enhanced by including feedback loops for evaluation by end users of technology and of the services provided by technology, taking the time dimension into account?

What are the characteristics of firms' problems that call for a technological solution?

How can measures include both positive and negative impacts?

Quantify and evaluate the effectiveness of government intervention, e.g. France Minitel vs US Videotex.

Evaluate the endogeneity/exogeneity of technology decisions, e.g. the endogeneity of decisions including regulatory decisions in reference to wireless vs wireline.

In what sense has an information revolution occurred similar to the industrial revolution of the 19th century?

The uneven trend towards modularization and greater scope for architectural innovation, as opposed to product and process innovation, needs further investigation. For example, the scope for architectural innovation in telecom services by non-dominant players, e.g. PTTs, is largely unexplored.

How do firms/households integrate new technologies into their processes? Are there good ways to do this?

To what extent is technological innovation in telecommunications geared to the reduction of labour cost as was the case in manufacturing?

How can the social usefulness of innovation rather than the mostly quantitative or material attributes of the products be measured?

What characteristics are required of telecommunications systems and products for their cost-effective operation in developing countries?

How can a two-way traffic in multimedia-based cross-cultural communications be secured?

4. Market structure

What factors influence the rate of turbulence in market shares in a high technology market? Is this turbulence good or bad?

How does telecommunications affect industrial structure?

Economics of "coopetition" (cooperation/competition and global alliances).

How can the needs of individuals be integrated into models of market structure?

Can the computer view of NII (as a concept) and the telephone cable view be accommodated theoretically in a process driven by consultation?

Can market structure, technology and regulation be analyzed as the outcome of endogenous processes, in order to explain and give prognoses for telecom industries' transformations over the decades from, say, the mid-1950s?

Many researchable issues arise because firm innovation is constrained by history, i.e. firm technological trajectories and divergences among firms within similar environments. Telecom service industries and users in general could be investigated in terms of:

- Sources of innovation, particularly the role of customers and the ability/inability of incumbents to innovate;
- The role of barriers to entry; and
- The network advantages of multi-locational vs single location firms and the distributional advantages of large vs small firms.

How do convergence and global alliances influence the appropriate structures?

What is the rationale for the faith in competition? Is there really competition in oligopoly situations?

To what extent is interoperability achieved in deregulated markets?

Price reduction, infrastructure and services development, etc. are reportedly associated with deregulation. To what extent can similar evidence be gathered in the case of monopolies, e.g. in France?

Measurement of concentration in the information industries and analysis of its consequences? What is the ownership structure of these industries?

Implications of dependence on foreign-operated and controlled communication systems.

5. Growth and productivity

There are spillovers for market structure, regulation and social dimensions. These need to enter into economic theory and regulatory activity — theory and practice — to control the rhetoric and advance knowledge.

Why do some countries do better/worse than other comparable countries?

What are the magnitudes of growth spinoffs from telecommunications investment?

How does network reliability influence productivity?

Given lessened dependence upon labour, what new measures of productivity are available?

Case studies of the efficiency of public administration.

Comparison of performance of SMEs with that of large, multinational firms.

National measures need to be complemented by regional, income group or firm size and urban/rural comparisons.

6. Regulation

Should we adopt a customer rather than a supplier approach to regulation? Who are the customers for regulation? What do they want of the regulator? How would we judge the regulator's success in meeting their needs?

What is the boundary of regulation, i.e. at what point does the regulator cease acting in the public interest and start acting as a surrogate telephone company manager?

How can we replace regulation with rules?

Political analysis of the term deregulation. What political theories underpin it? How does it manifest itself in different countries? What issues are obscured by government's acceptance of deregulation?

Exploration of the essentially political nature of markets and especially their institutional structures, e.g. when regulatory structures are transported to other countries.

Cross-country study of alternative regimes, in terms of structure and behaviour, including historical and cultural dimensions.

The two trajectories of regulatory sophistication and progressive competitive development. Are they inconsistent?

What institutional, economic and political factors affect the prospects for developing mechanisms and institutions for international telecommunications regulation?

Who/what controls resources and technologies as rapid changes manifest themselves, both within nations and across national boundaries?

How does regulation lower barriers? When does this stop?

How can we promote universal service and competition?

To what extent are the supposed requirements for regulatory bodies to be effective, e.g. institutional framework, staff expertise, availability of information, etc., met in less developed countries? How could they be met, or the limitations alleviated?

Is there a separation of power among government, industry and public interest groups when the players actually come from one community but can switch positions, e.g. a former Minister of telecommunications becoming consultant for an operator he previously permitted to enter the market?

How can the cost of universal access be shared among all operators?

7. Social dimensions

Does the growth of telecommunications and IT networking lead to more or less even regional development?

To what extent are transport and communications substitutes or complements?

In what ways do telecommunications affect power structures within society? What is the relationship between telecommunications and power? How will the convergence of telecommunications with broadcasting affect power and control structures in society?

To demonstrate the integral part that telecommunications and information play in all parts of life and segments of organizations and institutions.

There is a growing tendency, e.g. in economics, organization studies, management of technology, to treat space as non-homogeneous. This has particular relevance for communications technologies:

- Is the threshold size of firms being lowered and/or will the IT technologies favour large firms?
- What scope is there for separation and relocation of activities?
- What help does communication technology give with relocation/ adjustments?

If we agree that the *raison d'être* of telecommunications is not simply the result of the availability of technology and industry supply but their adequacy in meeting needs of individuals, groups and organizations to communicate, then social dimensions should have priority in research. The major issue is whether the provision of information and information facilities does or does not make a difference in abilities to cope with problems and to develop.

The potential of interdisciplinary research raises many issues:

- Which disciplines consider information as a central subject?
- Which disciplines are likely to offer relevant concepts?
- How are information phenomena described in the jargon of the various concerned disciplines? Can a unified language be framed?
- What subjects are currently treated in interdisciplinary research?
- Which priority subjects in telecommunications and information activities require, or are appropriate for, interdisciplinary work?
- How can an interdisciplinary approach be implemented?
- To what extent do the present administrative linkages of academia, government and industry bias, limit or inhibit the development of appropriate research and teaching?
- Is there a need for cooperative programs? Could/should they be applied to both theoretical and empirical studies?

- Have important contributions resulted from interdisciplinary work or were the landmarks contributed from a single discipline perspective?
- Is the establishment of the study of information and telecommunications as a particular discipline a better alternative?

References

Akerlof, G.A., 1984, *An Economic Theorist's Book of Tales* (Cambridge University Press, Cambridge).

Barlow, J.P., 1994, A taxonomy of information. *Bulletin of the American Society for Information Science* 20:13-17.

Brousseau, E., 1994, EDI and inter-firm relationships: Towards a standardization of coordination processes? *Information Economics and Policy* 6:319-347.

Carlson, S., 1979, *Swedish Industry Goes Abroad: As Essay on Industrialization and Internationalization* (Studentlitteratur, Lund).

Datapro, 1994, *Catalog of Global Telecommunications Reports* (Datapro Information Services Group,Sydney).

Economist, The, 1994, The tyranny of triangles. July 16, 65.

Geweke, J., 1984, Inference and causality in economic time series models, in: Z. Griliches and M.D. Intriligator, eds, *Handbook of Econometrics*, Vol. II (North-Holland,Amsterdam) 1101-1144.

Lamberton, D.M., ed., 1996, *The Economics of Communication and Information.* (Edward Elgar, Cheltenham, UK).

Marshall, Alfred, 1925, *Principles of Economics*, 8th ed. (Macmillan, London).

Media Information Australia, 1990, No.56.

Menou, Michel, ed., 1993, *Measuring the Impact of Information on Development* (IDRC Books, Ottawa).

Menou, Michel, 1994, From data to wisdom: Does IT contribute to gross national happiness? in: D.M. Lamberton, ed., *Beyond Competition: The Future of Telecommunications* (Elsevier, Amsterdam) 71-94.

Miller, George, 1983, Foreword, in: Fritz Machlup and Una Mansfield, eds, *The Study of Information: Interdisciplinary Messages* (Wiley, New York).

Phelps Brown, H., 1980, The radical reflections of an applied economist, *Banca Nazionale del Lavoro Review* 132.

Romer, Paul, 1994, New goods, old theory, and the welfare costs of trade restrictions. *Journal of Development Economics* 43:5-38.

Streeten, Paul, 1974. The limits of development research, *World Development* 2:11-34.

Taylor, Lester E., 1994, *Telecommunications Demand in Theory and Practice* (Kluwer Academic Publishers, Dordrecht).

Wall Street Journal, 1994, G-7 to hold telecommunications parley in the Fall at the suggestion of Clinton. July 7.

CHAPTER 2

The international economy, knowledge flows, and information activities

H.-J. Engelbrecht

Introduction

It is a pleasure and privilege to contribute to this research agenda. I shall concentrate on two broad avenues for research into the international dimensions of telecommunication and information activities which reflect my own current research interests as an applied economist. First, I suggest that a modified version of Antonelli's model of localized technological knowledge and network information processes (chapter 4) can be usefully applied at the international level. In fact, such an approach is necessary if one wants to shed light on the intriguing question of why inter-country differences exist in the levels of technological knowledge, the introduction and diffusion of innovations, and productivity growth. The question is also at the heart of the new theory of endogenous growth, which has been one of the most prolific areas of economic research in recent years.[1] However, only recently has work begun on empirical testing of what could be described as international percolation processes, in particular international R&D spillovers (Coe and Helpman, 1993; Mohnen, 1992; Bernstein and Mohnen, 1994; Irwin and Klenow, 1994). Although this work is promising and seems to have potentially major implications not only for our understanding of the process of economic growth in an international context, but also for industry, technology, telecommunications and a wide range of other government policies, it is only in its infancy and needs to be extended and refined.

Secondly, I shall suggest that the increasing identification over the last ten years or so of the "information economy" with the "information **technology** economy", the latter of which focuses mainly on the role of information and telecommunication technologies and is a reflection of the dramatic advances in these areas, has led to the relative neglect of the analysis of information activities in the wider sense. By information activities in the wider sense I mean all activities primarily associated with the creation, storage and dissemination of information. Such activities consume an increasingly large proportion of resources in any economy. I shall argue below why modelling of the wider information economy should be pursued, and suggest that incorporation of variables derived from an information, transaction cost and institutional economy perspective into applied research can provide valuable new insights into the role and effects of increased resource use for information activities.

Towards a model of international percolation processes and information networks

Country-specific technologies

The first task in developing the counterpart of Antonelli's firm model at the national level is to provide evidence that technology is in fact not only firm-specific but also country-specific, and that this is an important factor in explaining inter-country differences in economic performance. While nobody would deny the existence of institutional and cultural differences between countries, such differences and their relative importance compared to other factors are difficult to pin down empirically in mainstream economics.

Institutional economists, on the other hand, have long recognized the path-dependency of economic development, seeing institutions as carriers of history and as one of the ultimate sources of economic growth.[2] One of the more recent aspects of this view is the importance of "national systems of innovation" (Lundvall, 1992). However, mainstream economic journals have also recently reported that the data are at least consistent with important national differences in technology. Costello (1993), for example, presents evidence on the nature of productivity growth for five major industries in six of the major industrialized countries over the 1960-85 period. She finds that productivity growth is more correlated across industries within one country than across countries within one industry (*ibid.*:207). A substantial fraction of changes in productivity can therefore be attributed to country-specific factors that are common across industries.

There are many recent studies dealing with certain aspects of the world economy as an information percolation network. In the following, I shall concentrate on studies which focus on either the emission aspect (quality of connectivity) or reception aspect (quality of receptivity) of the communication process.

Connectivity: the Coe and Helpman study

Coe and Helpman (1993) is an example of the former. Their model is based on recent theories of innovation-driven growth of the expanding product variety type (where R&D leads to the invention of new products) and the rising product quality type (where R&D leads to improved quality of products) models (see Grossman and Helpman, 1991, chapters 3 and 4). These theories treat commercially-oriented innovation efforts which respond to economic incentives as a major engine of technological change and productivity growth.

Coe and Helpman's data sample consists of 21 OECD economies plus Israel during the period 1970-1990. The data are for the business sector only, i.e. business sector R&D, not total R&D, is used. The novel feature of their model is that total factor productivity (TFP) is made to depend not only on accumulated domestic R&D capital stocks, but also on accumulated foreign capital stocks. The former is already a well-established determinant of domestic TFP growth.[3] Their basic estimation equation in general terms is as follows:

$$\text{TFP} = f(\text{RD}^{\text{D}}, m^{*}\text{RD}^{\text{F}}) \tag{1}$$

where RD^{D} represents the domestic R&D capital stock, which is taken to be a proxy for a country's stock of technical knowledge. RD^{F} is the import weighted foreign R&D capital stock, and m is the fraction of imports in GDP. Foreign R&D capital stocks are scaled by m to reflect the assumption that a country that imports more relative to its GDP should experience larger TFP benefits from foreign R&D. The data also support the view that the impact of domestic (but not foreign) R&D on TFP differs between small and large countries. Therefore, Coe and Helpman include a $\text{G7}^{*}\text{RD}^{\text{D}}$ term in their preferred equation (G7 is a G7 dummy variable).

Their estimates indicate that foreign R&D has beneficial effects on domestic TFP, and that these may be stronger the more open an economy is to international trade. Some interesting inter-country differences emerge, for example Australia has a low spillover elasticity, i.e. that country does not appear to be able to take much advantage of overseas knowledge. Foreign R&D has a larger impact on TFP than does domestic R&D for most of the smaller economies (see Table 1). Table 2 provides an impression of the international knowledge diffusion process. It can be seen that the domestic R&D capital stocks of the US and Japan create the largest international spillovers. Coe and Helpman also find that rates of return on R&D capital stocks are very high, both in terms of domestic output and international spillovers.

Policy implications

The policy implications of these findings may be profound. While the non-rival nature of disembodied knowledge implies that the private sector is likely

Table 1. Country-specific, time-varying estimates of the impact of R&D capital stocks on total factor productivity

	Elasticity of total factor productivity with respect to:				
	World R&D	Foreign R&D			Domestic R&D
	1990	1971	1980	1990	1971-90
United States	0.267	0.016	0.030	0.033	
Japan	0.261	0.028	0.037	0.027	
West Germany	0.311	0.056	0.072	0.077	
France	0.301	0.045	0.061	0.067	0.234
Italy	0.292	0.046	0.067	0.058	
United Kingdom	0.315	0.063	0.081	0.081	
Canada	0.309	0.059	0.078	0.075	
Australia	0.132	0.043	0.049	0.055	
Austria	0.192	0.091	0.106	0.114	
Belgium	0.337	0.129	0.181	0.260	
Denmark	0.169	0.091	0.094	0.092	
Finland	0.152	0.079	0.088	0.075	
Greece	0.172	0.050	0.063	0.094	
Ireland	0.243	0.124	0.180	0.165	0.078
Israel	0.231	0.147	0.154	0.153	
Netherlands	0.236	0.133	0.146	0.158	
New Zealand	0.144	0.075	0.086	0.066	
Norway	0.188	0.133	0.124	0.111	
Portugal	0.210	0.099	0.117	0.132	
Spain	0.141	0.043	0.043	0.063	
Sweden	0.171	0.067	0.087	0.093	
Switzerland	0.190	0.115	0.106	0.113	

Source: Coe and Helpman (1995), Table 4, p.15.

to undersupply research activity which might justify government policies in support of R&D,[4] Coe and Helpman's findings weaken the case for support of "strategic industries" because subsidizing domestic industrial research and development also indirectly subsidizes foreign firms. Spillovers cannot be kept within national boundaries.

Coe and Helpman's findings also strengthen the argument in favour of free trade in goods and services and against government intervention in this area. Helliwell and Chung (1991), for example, find significant evidence that technical progress has been faster, *ceteris paribus*, for countries that increased their openness to international trade. Coe and Helpman's findings provide one explanation for this observation. Moreover, Romer (1994) argues that economists, especially those trained in formal mathematics and general

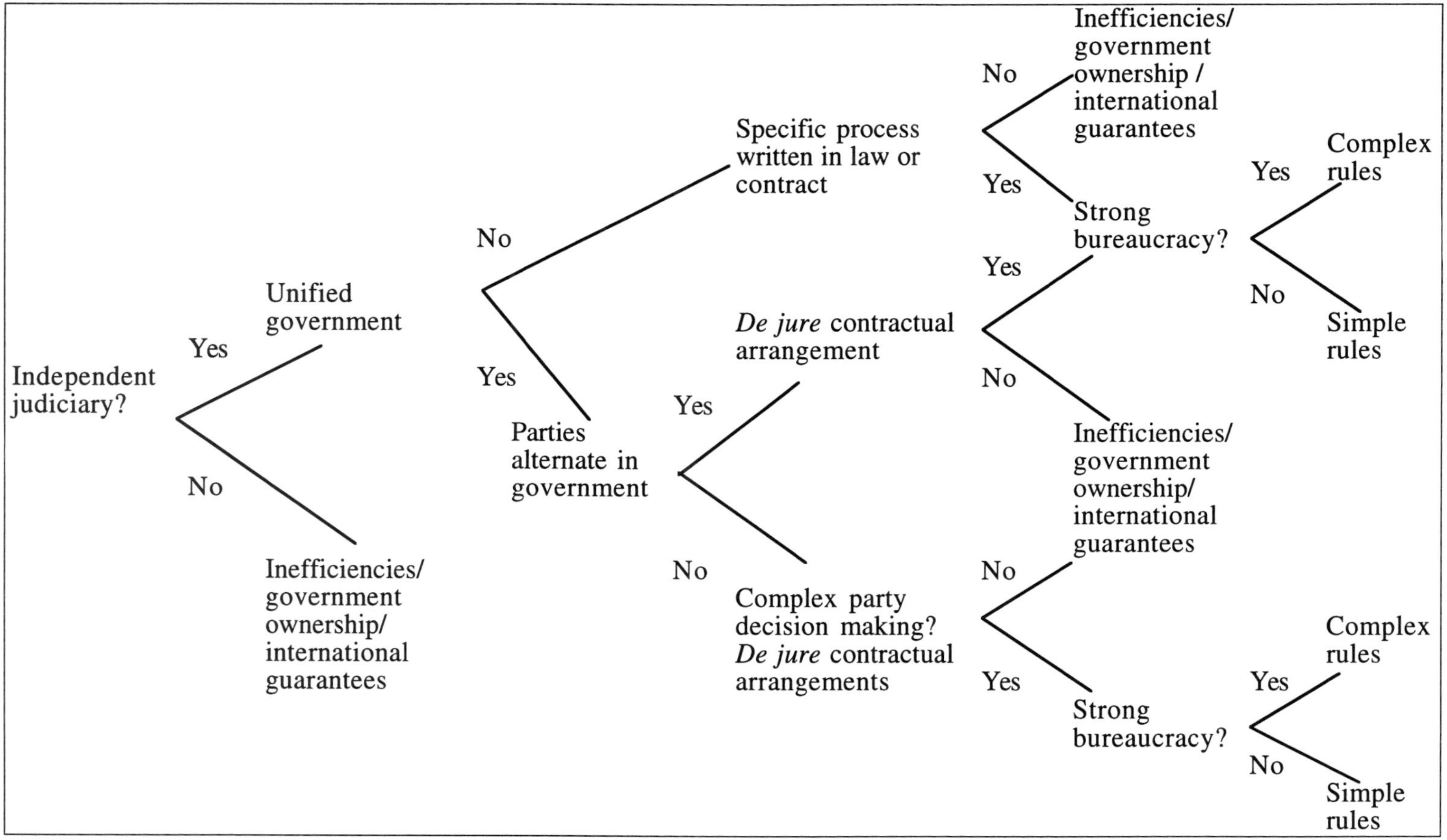

Figure 1. A decision tree model of a country's institutional endowments.

equilibrium theory, have greatly underestimated the welfare costs of trade restrictions and interferences precisely because they have neglected the dynamic effects such policies have on the innovation potential of an economy:

> The true costs of badly designed government interventions, and especially of trade restrictions in developing countries, come not from their effects on the static allocation of resources between the activities in an economy that already exist. Rather, they come from the stifling effect that the distortions have on the adoption of new technologies, the provision of new types of services, the exploitation of new productive activities, and on imports of new types of capital goods and produced inputs (*ibid.*:25).

Romer sees a more fundamental reason than mathematical difficulties for why the implications of new goods have not been more extensively explored by economists in the past in a deep philosophical resistance humans feel, going as far back as Plato, towards a world view that assumes new things can happen and could have happened at every date in the past.

A research agenda for international spillover studies

While Coe and Helpman's study provides a snapshot of the global information percolation processes, there are many weaknesses and shortcomings in their approach which point to a wide research agenda:

1. They neglect some important other variables suggested by the new growth theories which, when included, should reduce the impact of domestic and foreign R&D on TFP. In particular, human capital has been at the centre of one branch of the new theories (e.g. Lucas, 1988, 1993). Its neglect by Coe and Helpman might explain their unrealistically large estimates of rates of return on R&D. Also, it might well be that development of (general) human capital has a stronger effect on TFP in some countries compared to more specialized investment in R&D capital. What are the trade-offs? What is the critical mass of "R&D human capital" in smaller countries needed to sustain or increase the capacity to absorb international technology spillovers ("absorptive capacity")?
2. In terms of model building, the use of a single equation approach as used in new growth theories does not seem appropriate. Significant feedbacks between R&D and TFP, and other variables included in the different models, require a simultaneous equation approach (Fagerberg, 1994). This should help overcome the problem that a given pattern of correlation in macro data of the type used in support of new growth

theories can be due to many different causal structures, i.e. the same data usually support different, and competing, theories.

3. The use of trade weighted R&D capital stocks implies that either all technology flows are embodied in traded intermediate goods and services or that such commodities are an acceptable proxy for the many other mechanisms by which knowledge is transferred between countries. Such other mechanism include:

   i) multinational companies and foreign direct investment, ii) joint ventures and licensing agreements, iii) business trips and international conference attendance, iv) education and training received abroad, v) migration of skilled people, vi) importation of journals and books, vii) purchase of patents and trademarks, viii) the role of trade associations as agents facilitating the dissemination of information, ix) reverse engineering (in which case the level of imports is unrelated to the level of knowledge transfer), x) last but not least the role of electronic information transfers (phone, fax, the internet).

   Romer (1993a) has recently rejected the premise of his earlier work that capital and ideas move in lockstep, implying that all information transfer mechanisms should be measured and analyzed, if possible. This might also help overcome the problem of assessing causation mentioned above. Romer (1993a:558-559) comments that his earlier work, in which he interpreted a large coefficient on physical capital to be accompanied by investments in new ideas, might have done more harm than good for the development of new growth theory, and that it was an act of desperation dictated by the lack of direct evidence on technology and ideas.

4. The empirical measurement of TFP, R&D, human capital, methods of knowledge transfers etc., insofar as it exists, needs to be greatly improved before more reliable conclusions can be drawn.[5] For example, how good a proxy are domestic R&D stocks for the level of domestic technological knowledge? Should patent statistics also be used (or instead)? How useful are the new trade literature based innovation accounts (e.g. Kleinknecht and Bain, 1993)? Some of these issues were taken up in more detail by Swann in chapter 5. Is there any realistic chance to investigate the timing problem, i.e. R&D and spillover lags?[6] What are the different lags for different modes of knowledge transfer? Would it be more appropriate to use a subgroup of imports as weights for overseas knowledge stocks, e.g. capital goods imports, instead of all goods and services as in Coe and Helpman?[7]

5. There is also a need to focus on the effects of developments in telecommunication and information technologies. How will they affect the

relative importance of the different modes of knowledge transfer? How will they affect their lag structures? Will they enhance or reduce the appropriateness of R&D as a proxy variable? For example, how much will knowledge spillovers be speeded up and strengthened, duplicate research be reduced and absorptive capacity increased, if patent offices put their databases on the internet, or if governments made research funded with public money available on the internet, as is already the case with European Union funded pre-competitive research?

6. A related issue is that of the effects the new telecommunication and information technologies have on regional development (at the sub-national level). As pointed out by Swann (chapter 5) and Maddock (chapter 7), we need more evidence and further theoretical work in this area. In the context of international knowledge flows, we have to ask what the impact of such flows is likely to be on regional disparities, i.e. will they reinforce or erode regional differences? Are some modes of knowledge transfer more suited than others to stimulate regional development?

Some other studies

So far there are few other cross-country studies of spillover effects, be it at the firm, industry or national level. Irwin and Klenow (1994) analyze the semiconductor industry and find that spillovers in this "strategic" high-tech industry are international in scope: learning spills over just as much between firms within a country as it does between countries. They therefore do not provide a justification for policies that favour domestic over foreign firms. Mohnen (1992) investigates the role of foreign, i.e. mainly US, R&D spillovers in Canadian manufacturing during 1964-1983. His findings seem to indicate that foreign R&D is complementary to domestic R&D. Bernstein and Mohnen (1994) estimate Japan-US spillovers in R&D intensive ("high-tech") industries and find them to be complementary to domestic R&D intensity in the US in both the short and long-run, but only complementary to domestic R&D in Japan in the short-run. International R&D spillovers contribute to productivity growth in both countries. Both countries, but especially Japan, also decrease their labour intensity in response to spillovers.

Receptivity

Parente and Prescott (1994) is an example of a recent study focusing on the receiving end of the communication process. Their theory focuses on the role of technology adoption and barriers to it. It is an attempt to develop a theory of economic development which can account for both the huge income disparities between countries and development miracles.[8] All firms in every country are assumed to have access to world knowledge, but each country is assumed to have a different level of barriers to technology adoption, i.e. regulations and legal constraints, the institutional environment in general. The larger these barriers, the greater the investment a firm must make to adopt a more advanced technology. Large persistent reductions in these barriers induce development miracles. They calibrate their model on US and Japanese data and show it to be quantitatively consistent both with the great income disparities between countries and development miracles. Similarly to Lucas (1993), they conclude that there must be a large unmeasured investment in the business sector, which they view as technology adoption investment in contrast to Lucas's learning on the job.

However, while Parente and Prescott's story seems plausible, their cost estimates are obtained implicitly. In a sense they are a measure of their ignorance about other development factors. As they point out (1994:319), the task ahead is to obtain direct measures of the magnitudes of barriers to technology adoption, or cost of learning/absorptive capacity, and empirically test their relevance. Secondly, we need to understand why such costs might differ between countries. Similarly to Coe and Helpman, they suggest that trade openness contributes to development, in this case because it weakens the forces of resistance to technology adoption.

Receptivity supporting policies

The question Parente and Prescott have not addressed is that of the design of policies which will overcome country-specific "adoption" problems and ensure continuing low resistance to technology adoption. What is the government's role in the information world economy in this regard? What policies can be devised that support information flows and the absorptive capacity of nations, i.e. that increase the percolation probability of the system?

Given the network properties described by Antonelli in chapter 4, the answers might not so much depend on the availability of the latest information and communication technologies (i.e. connectivity), but on the whole range of government policies, i.e. trade, industrial, telecommunications, labour market and training, foreign investment, social and cultural policies, as well as the institutional set-up, which all affect receptivity to new information and flexibility in regard to structural change in the economic system. To a certain degree, most government policies are "information policies in disguise", in the sense that they have mainly other objectives but major impacts on information flows and the quality of receptivity.

I do not want to minimize the importance of connectivity, i.e. the effects telecommunication technologies have on the process and the diffusion of technological knowledge and economic development in general (the reader is referred to the discussion of this topic by Maddock in chapter 7), but it is important to highlight the wider policy and institutional context. Let me give some examples. Romer (1993a:565) argues that the correlation between poor macroeconomic policy, the inflation rate and economic growth found by Fischer (1993) is easier to understand if we assume that such policies reduce the in-flow of ideas from abroad, for example by reducing the involvement of multinational companies. Another example is provided by Weder and Grubel (1993), who show how a country's private institutions, in particular voluntary industry associations, company structures and industry clusters, can internalize knowledge spillovers. Similarly, Romer (1993b) advocates the creation of self-organizing industry boards which would have the right to petition government to impose a levy on a particular industry grouping in order to fund public good innovations and facilitate information flows amongst member firms. There is also the intriguing question, discussed at some length by Swann in chapter 5, of how the level of intellectual property protection might affect networking, and whether it would be optimal to give preference to the latter over the former. A related issue is that of putting restrictions on information flows in networks in order to preserve competition and counter collusion (legal versus illegal information flows). Where are the boundaries? Can they be established and agreed upon at the international level? Do we have the knowledge to distinguish between competitive and pre-competitive technology flows, or "fair" and "unfair" network arrangements? To give one pertinent example, should information flows within Japanese Keiretsu, i.e. the groups of sub-contracting firms arranged around a core manufacturer, be seen as a positive example of an institutional arrangement which internalizes knowledge spillovers, or as an unfair competitive advantage?

The task ahead: simultaneous modelling of receptivity and connectivity

Coe and Helpman try to measure the effect of international R&D spillovers. Parente and Prescott try to measure the effects of adoption barriers, while assuming that production knowledge spills over fully and instantaneously. Either theory focuses on only one side of the communication process, i.e. either the emission of information or the reception of it. An interpretation of the world economy in network economics terms requires that both aspects are modelled simultaneously. We need to know what the spillovers are, but also the extent to which they are able to be taken up. The creation and transmission of technological knowledge in the world economy depends on the percolation probability of the system, to use Antonelli's description, i.e. it depends on the quality of the information and communication channels (the quality of connectivity) and the degree of absorptive capacity (the quality of receptivity). Moreover, the latter is both a determinant as well as a property of localized technological knowledge. Antonelli also suggests that receptivity matters more than connectivity, and that it is more effective to try to modify the former compared to the latter.

One simple extension of a Coe and Helpman type model that immediately comes to mind is to add an interaction term between the spillover variable and domestic R&D. This might provide some quantitative indication of what proportion of domestic R&D capital is needed to absorb foreign knowledge. On the other hand, international knowledge spillovers will be a substitute for some domestic R&D activities. What is the right balance between conducting own research, both in order to develop new products and in order to strengthen absorptive capacity, and the level of imported technology? How does the balance between receptivity and connectivity differ between countries? Given limited resources, what is the optimal trade-off? How is this affected by domestic institutional, social, political and cultural factors? For example, economic historians have pointed out the importance of social resistance to technological progress, and the concomitant anti-technology ideology permeating the education system and labour relations, as an explanation for the loss of technological leadership of Victorian Britain (Mokyr, 1992).

While researchers should strive to improve the measurement of information flows, we have to agree with Romer (1993a) that it is unlikely that all of them can be measured in quantitative terms. In order to understand the significance of "idea gaps" for economic development which can be overcome by information flows, he argues that economists should use evidence from both mainstream formal models and the diverse evidence of what Nelson and Winter (1982) call "appreciative theory", i.e. the less formal evidence provided by economic historians, analysts of technological change, institutional economists, New Keynesians and other intellectual traditions outside mainstream economics which might help us to pin down the important causal relationships and feedback mechanisms.

Absorptive capacity as capital

The absorptive capacity for international knowledge flows associated with domestic R&D can also be interpreted as a form of capital, which can depreciate and become obsolete. Particularly in cases of major technological breakthroughs (i.e. radical innovations or technological revolutions), it is likely that an increase in absorptive capacity to take advantage of the new technology depreciates the usefulness of existing adaptive capital, i.e. there is Schumpeterian informational creative destruction (Jones and Newman, 1995) of localized technological knowledge, including tacit knowledge. One implication of this view is that a high level of existing adaptive capital does not always ensure low adoption costs of new technology, or a technology adoption advantage over countries having a lower level of adaptive capital. However, a high level of adaptive capital will ensure that improvements along an existing technological trajectory (i.e. incremental innovation) will be taken up relatively quickly, i.e. in most cases one can expect a strong hysteresis effect in the technology specialization of countries. However, as pointed out by Swann (chapter 5), it is very difficult to distinguish between incremental and radical innovations *ex ante* — or even *ex post,* i.e. it is equally difficult to forecast and assess the rate of depreciation of absorptive capital.

If it is accepted that investment in absorptive capacity (adaptive knowledge capital) is technology specific, then what appears to be a low absorptive capacity for a particular technology might be explained as a rational way of avoiding the destruction of previously accumulated adaptive capital. How could the importance of this historical lock-in effect be tested empirically? Could it be a possible explanation for Australia's seemingly low capacity to take advantage of foreign technology?

Related concepts

There are a number of related concepts in the literature which try to capture "absorptive capacity". The concept of "social capability" has been used to capture those country characteristics which influence the degree of technology adoption, especially social and political factors (see, e.g. Hansson and Henrekson, 1994, Fagerberg, 1994, and the references cited therein). It has tended to include so many different factors that at a conceptual level it seems to be little more than a catch-all term for country-specific effects. In practice, the proxies used have often made it indistinguishable from other theories.[9] The concept of "national systems of innovation" covers a similarly wide variety of national factors and therefore can be similarly criticized.[10] A related concept, and one theoretically subsumed under both social capability and national system of innovation, is that of "organizational capital" (Tomer, 1987). However, in practise it is distinct enough to warrant separate discussion.

Information activities and organizational capital

A pertinent distinction neglected so far is that between technological knowledge and technological spillovers on the one hand, and organizational knowledge and factors that might lead to organizational change on the other hand. While economists usually acknowledge that the right organizational capabilities are needed and indeed are essential for successful adoption of new (production and communication) technologies in both developed and developing countries (see Maddock, chapter 7, Freeman and Soete, 1990, Mody and Dahlman, 1992), analysis of "organizational technology" is usually excluded, or put in the too hard basket (see Swann, chapter 5). As Matthews (1986) has pointed out, technical efficiency in the production sector seems easier to achieve than advances in organizational efficiency.

I suggest that another reason why questions concerning organizational efficiency have received relatively little attention in the mainstream economics literature is because the resources devoted to the organizational task are not normally explicitly taken into account in economic analysis. For example, in his survey article Radner (1992:1383) finds it necessary to first provide empirical evidence of the size of the "management sector" in the economy in order to "convince readers that managing is an activity worthy of economic analysis".

In the following I shall not discuss the economic theories which have been developed to explain and analyze different organizational structures. The interested reader is referred to Radner. Instead, I want to make a case for a different way of analysing organizations, and other information based activities. It concerns measuring, and than analysing, the costs of information activities. In short, I want to direct discussion towards information economics, or the analysis of the information economy, a topic which has been surprisingly absent from this workshop. It is my contention that we can say a lot more about the effects of resource use for informational purposes than we have done so far.

The information economy

At the economy-wide level, changes in economic structure towards information-based economies have been measured and analyzed in the pioneering studies of Machlup (1962), Bell (1973) and Porat and Rubin (1977). The latter gave rise to a series of similar studies, stimulated by work on the measurement of the information economy by the OECD.[11] These studies aim to explicitly account for all goods and services associated with the creation, storage and dissemination of information, where information is defined in a restricted sense: only information relating to production as measured in the national accounts is included. Many information goods and services are exchanged in established markets, others are non-marketed. Therefore, the information economy is divided into two components, i.e. the primary (PRIS) and secondary (SIS) information sector. The PRIS comprises, first, goods and services which intrinsically convey information or which are directly useful in its production, processing and distribution and which, secondly, are transacted on established markets. However, the PRIS does not account for resources devoted to private bureaucracies which produce and consume information services "in-house". These non-marketed services are ancillary to the production of non-PRIS goods and services and are integrated into a consistent input-output framework. To measure the SIS, non-PRIS firms are taken apart in an accounting sense into an information division (bureaucracy) and non-information division (production). These firm divisions, or "quasi-firms" (Porat and Rubin, 1977), are aggregated to the sector level, to make up the SIS and non-information sector of the economy respectively. The data requirements for the measurement of the SIS in particular are formidable (see, for example, Engelbrecht, 1986).

However, since the mid 1980s, the OECD has shifted its work towards other aspects of the information society, focusing on more narrowly information technology related issues (see OECD, 1993, and other volumes in the Information, Computer, Communications Policy series). This also applies to many other information economy analysts (see, for example, Miles, 1990; Imai, 1990).

The transaction economy

On the other hand, the importance and relevance of research on measurement and analysis of the wider information economy is underlined by developments in transaction cost economics. The latter is closely related to information economics because many in-house information activities are also transaction activities. Because such costs now tend to dominate overall costs, productivity growth and competitive advantage increasingly depend on economizing on the use of resources devoted to these activities.

There are very few attempts to measure transaction costs at the economy-wide level. Wallis and North (1986) is one prominent example. Although information and transaction costs are not identical (see, for example, Eggertsson, 1990, chapter 1), in terms of empirical measurement they are very similar, in particular in regard to the "in-house" sector. Like Porat and Rubin, Wallis and North employ two measures. First, certain economic activities are classified as "transaction industries", secondly, there are transaction services that occur within firms of "non-transaction industries". The latter are measured by dividing occupations into those primarily providing transaction services and "others". By far the largest groups making up this sector are managers, clerical and sales workers, i.e this sector overlaps to a large extent with private bureaucracies or the management sector.

Problems with both approaches

The problems with this approach are similar to those associated with the measurement of the information economy. First, measurement is sensitive to the definition of transaction or information occupations.[12] While there are some tricky issues involved in delineating such occupations, this is however not impossible. Secondly, there is the problem that the in-house sector of the PRIS, and that of transaction industries, is not explicitly measured. This criticism goes back to Machlup (1980), who insisted that PRIS and SIS are two distinct approaches which should be strictly kept apart. It has also been acknowledged by Rubin (1988). The SIS should be measured for every industry. This would also provide advantages in terms of empirical measurement, i.e. one would not have to first measure PRIS or transaction industries from officially published data before beginning measurement of the SIS (Engelbrecht, 1989/90). An example is Rabeau (1990), who estimates Canada's information economy by information worker value added in all sectors of the economy.

The problem then is that such an in-house sector not only includes information and transaction activities used for organizational purposes, but also for production, i.e. in some industries information (and transaction) activities are not ancillary to the main productive activity, but part of it. For many purposes it might well be desirable to distinguish which in-house information or transaction activities should be regarded as part of "production" instead of "organization", a problem also acknowledged during the workshop by Hagström, who gave the example of telecom use in software production (see also Carter, 1989). While the measurement issues are difficult and controversial, I do not think we should give up on them.

Further analyses

I argue that the relative neglect of the measurement of resource use for information activities other than those related to information technology has pushed some of the most interesting questions concerning information activities onto the sideline.

Information resource use data generated by information economy studies can be analyzed further, for example they have been incorporated in production function analysis (Hayes and Erickson, 1982; Warskett, 1984; Braunstein, 1988; Engelbrecht 1987/88, 1990) and general equilibrium models (Engelbrecht, 1989/90; Karunaratne, 1991). An information economy has yet to be incorporated into a Social Accounting Matrix. In principle, the latter is extremely flexible and can incorporate much additional information, e.g. on the distribution of income, which would enable analysts to address a wide variety of policy problems associated with the information economy. Also, it might be possible to include household activities which are not part of the formal economy. We might then be able to analyze how this large part of the "unrecorded" economy is affected by information activities and flows. It should also become possible in future to combine information technology usage indicators which are being collected by the OECD (1993) with information economy studies. This should provide further insights into the relationship between information technology use and information activities in the wider sense.

The data generated by information and transaction economy studies have also been incorporated into productivity analysis (e.g. Porat and Rubin, 1977). Fuess and van den Berg (1992) find that growth of transaction industries has thwarted US productivity growth, especially during the 1980s. The transaction industries could be analyzed in detail. For example, as social scientists concerned with information activities we should be researching the importance and effects of such characteristics of the US information economy as the rise in the number of lawyers.

There is also a diverse literature, mainly by business, post-Keynesian and other "alternative" economists, which provides evidence that growth of the in-house sector might have had a similarly negative effect to that of transaction industries (e.g. Alexander, 1992). It should be analyzed to what extent information and transaction costs have increased over time without increasing output in the economy. Has growth of the "in-house" sector therefore distorted standard of living estimates? What effect might information and telecommunication technologies have on the level and/or change of these costs? On the last point, see Norton (1992), who also surveys the few existing studies in this area. The literature on the relationship between telecommunications and economic growth in general is discussed in detail by Maddock in chapter 7.

As pointed out by Williamson (1993), applications of transaction cost economics to country studies, that is the comparative study of economic systems and their performance, are just now being considered and much remains to be done. The need for comparative studies is also highlighted by Castells and Aoyama (1994), who find that the information economies of G7 countries display quite distinct patterns of evolution.

Technological change and organizational inefficiency: an example

In an age where most firm activities consist of information generation and processing, and information (and transaction) costs dominate total production costs, the explicit measurement and modelling of private bureaucracies and organizational capital can no longer be neglected.[13] Like absorptive capital, organizational capital can depreciate and become obsolete. Informational and organizational efficiency have become of paramount importance. A recent example is the rise in nonproduction employment relative to production employment in US manufacturing during the 1980s, which (unlike during earlier periods) was accompanied by a rise in nonproduction to production workers wages.

There seem to be two major alternative, but not mutually exclusive, interpretations of these changes. On the one hand, several studies suggest that they are due to technological change in favour of nonproduction workers, particularly due to the increased use of computers (e.g. Krueger, 1993; Berman *et al.*, 1994). On the other hand, Caves and Krepps (1993) have argued that they are due to organizational factors which resulted in over-employment, i.e. organizational "fat" and low organizational efficiency, concepts abhorred by neoclassical economists because they require abandonment of the view of the firm as a unitary profit-maximizing entity defined by a technologically determined production function.[14]

One approach to investigate this problem is to include an organizational variable in otherwise conventional applied trade studies. For example, Engelbrecht (1991, 1994, 1995b) tests the hypothesis that private bureaucracies are a trade determinant, and that relatively large private bureaucracies are a sign of organizational inefficiency, having a negative impact on productivity and on export competitiveness in manufactures. The estimation results confirm the hypothesis. They also indicate that the "organizational factor" is at least as important, if not more so, than traditional trade determinants, e.g. human and physical capital. It seems that both technological change and the lack of organizational innovation were responsible for the labour market developments in the US during the 1980s. This also provides a plausible explanation for the large increase in white-collar unemployment during the recession beginning in mid-1989, and the continuing corporate downsizing. The recession increased competitive pressure to such a degree that corporate restructuring and organizational innovation could no longer be postponed and the institutional rigidities acting against attempts to increase organizational efficiency were overcome (see also Engelbrecht, 1995a).[15] However, the organizational variables used in these studies are very simple and proxy a variety of organizational features. Future empirical research should be directed towards the explicit incorporation of organizational adaptation and learning, and of the more detailed determinants or organizational efficiency at the firm and industry level, for example subcontracting and networking structures.

The findings also indicate that low organizational efficiency can strongly counteract the positive contribution of R&D, (general) human capital, and, by implication, that of new information and communication technologies, to trade performance and growth. These interactions need careful modelling. A single equation approach again seems to be too simplistic, neglecting important causal relationships between the various determinants. What is the relative importance of technological capability and absorptive capacity in the narrow sense *vis-à-vis* organizational capability? How do they interact? These are questions which are also at the heart of the evolutionary economics research agenda (see the recent survey by Nelson, 1995). Again, what is needed is formal theorizing guided by a strong dose of appreciative theory.

Policies for organizational efficiency

The question whether a separate information policy is needed to enhance organizational capability and efficiency might be misleading if it diverts attention away from all the other policy areas which affect the receptivity to new information. As pointed out above, the thrust of a multitude of government policies has to be redirected to that aim. This is not to deny the desirability of devising specific government policies aimed at increasing organizational and informational efficiency, and organizational learning, of a nation's firms. Tomer (1993), for example, tries to develop a new rationale for government intervention based on this view. He concludes that a combination of infant industry argument and the concept of organizational learning can indeed provide such a rationale. He also argues that Japanese industrial policy is loosely based on this concept. The case of Harley Davidson, the motorcycle manufacturer, is given as an example of successful US government intervention in order to promote organizational learning (*ibid.*:217-219).[16] However, given the practical difficulties in identifying the special circumstances under which government intervention might be appropriate and successful, and the general benefits associated with free trade, I would expect such intervention to be the exception, rather than the norm.

Concluding comment

While the dual role of R&D as the generator of new information and facilitator in the absorption of existing information has long been recognized by historians of technological change, it has only fairly recently found its way into the industrial organization literature and the new growth theories.[17] As indicated above, it has yet to be taken up fully at the international level of analysis. The interpretation of the world economy as an information percolation network suggests a lengthy research agenda, as does further analysis of the "information economy". The overall aim should be to increase the percolation probability of the system. In particular, economic success depends on informational efficiency, at the same time as new technologies provide important opportunities to increase the connectivity of the information networks. Of course, all of these factors are interrelated, and some of the most interesting research questions emerge at their interface. How do new information and communications technologies affect absorptive capacity in the narrow sense, and organizational capacity of the "in-house" sector, and how does the lock-in of organizational capital affect the successful adoption of the new technologies? How can organizational efficiency and organizational change be measured? How could measures of organizational and absorptive capital be modelled alongside each other in an analysis of international knowledge transfers?

Notes

1. There have been many recent surveys and symposiums on new growth theories. See, for example, *Journal of Economic Perspectives* (1994), *Journal of Monetary Economics* (1993), *Economic Journal* (1992), *Oxford Review of Economic Policy* (1992), Amable (1994), Rebelo (1992). The seminal paper by Romer (1986) is usually regarded as the beginning of endogenous growth theory.
2. See, for example, *Structural Change and Economic Dynamics* (1994), North (1993), *World Development* (1989). Eggertson (1990) and Furubotn and Richter (1992) are excellent surveys of modern institutional analysis.
3. For recent surveys see, for example, Nadiri (1993) and Griliches (1992). However, while there have been many studies of the effects of R&D on output or productivity at the firm and industry level, there is a lack of similar cross-country studies. One of the first attempts is Lichtenberg (1993).
4. Moreover, non-rivalry of knowledge may not be a good enough reason for government actively supporting industrial R&D. If spillovers from learning-by-doing are more important than R&D as such, the scale of production might matter more (Young, 1993).
5. On TFP and R&D measurement see Griliches (1992, 1994); on human capital measurement see, for example, Behrman and Rosenzweig (1994) and Gundlach (1994). Increasing research on telecommunication and computer networking traffic statistics should provide new opportunities to analyze international information flows (see Lamberton, 1994:45-46).
6. For example, Adams (1990) finds that academic knowledge creation, which is not included in Coe and Helpman's R&D stock measures, strongly contributes to growth, but that lags are exceptionally long, i.e. in the order of 20 years from the appearance of research in the science community to its peak effect on production.

7. Mohnen (1992), for example, uses "high-tech" products as weights. He interprets such imports not as carriers of foreign R&D, but as a proximity measure in international technology space. This is closer in spirit to many of the firm and industry level spillover studies, where technological proximity plays a major role.
8. It is therefore in direct competition with Lucas's (1993) model based on human capital accumulation.
9. For example, Hansson and Henrekson (1994) use a human capital variable and trade openness as proxies. These variables have also been used in many other growth theories.
10. See, for example, Lundvall (1992) and McKelvey (1991). The latter contrasts and evaluates a number of different conceptions of national systems of innovation.
11. See, for example, OECD (1981, 1986), Jussawalla *et al.* (1988). For an alternative definition of the information economy see, for example, Jonscher (1983). Don Lamberton has surveyed developments in this research area on many occasions.
12. See Davis's (1986) comments on Wallis and North's work, and Schement's (1990) comments on Porat and Rubin's estimates of the information labour force.
13. Such neglect in fact distorts our interpretation of intersectoral relationships and structural changes taking place in the economy (Engelbrecht, 1992).
14. For more details, see Engelbrecht (1995b).
15. It would be interesting to conduct an indepth study of the changes in the US SIS for the 1980s and 1990s and compare them to changes during the Great Depression which, according to Porat and Rubin's (1977:166-168) estimates, was the only period during 1929 and 1974 (the end-point of their study) during which the SIS contracted.
16. While a mature firm, Harley Davidson's organizational capital had depreciated to such an extent that in organizational terms it could be seen as an "infant" which required support in the form of temporary protection.
17. See, for example, Cohen and Levinthal (1989) and the sources quoted therein. The question of absorptive capacity of firms is increasingly being taken up by industrial economists. At the 1995 Australian Industrial Economics Conference, Australian National University, which took place parallel to this workshop and which provided the author with many opportunities to take advantage of knowledge spillovers, absorptive capacity was one of the dominant reoccurring themes. Geroski's (1994) contribution warrants particular mention.

References

Adams, James D., 1990, Fundamental stocks of knowledge and productivity growth, *Journal of Political Economy* 98(4):673-702.

Alexander, David, 1992, Resource use and US manufacturing productivity growth, *Journal of Post Keynesian Economics* 14(3):389-407.

Amable, Bruno, 1994, Endogenous growth theory, convergence and divergence, in: Gerald Silverberg and Luc Soete, eds, *The Economics of Growth and Technical Change: Technologies, Nations, Agents*, chapter 3 (Edward Elgar, Aldershot) 20-44.

Behrman, Jere R. and Mark R. Rosenzweig, 1994, Caveat emptor: cross-country data on education and the labor force, *Journal of Development Economics* 44(1):147-171.

Bell, Daniel, 1973, *The Coming of Post-Industrial Society* (Basic Books, New York).

Berman, Eli, John Bound and Zvi Griliches, 1994, Changes in the demand for skilled labor within U.S. manufacturing industries: evidence from the annual survey of manufactures, *Quarterly Journal of Economics* CIX:367-397.

Bernstein, J. and P. Mohnen, 1994, International R&D spillovers between US and Japanese R&D intensive sectors, NBER Working Paper 4682 (Cambridge, MA).

Braunstein, Yale M., 1988, Information inputs and outputs in firms and quasi-firms, in: Meheroo Jussawalla, Donald M. Lamberton and Neil D. Karunaratne, eds, *The Cost of Thinking: Information Economies of Ten Pacific Countries*, chapter 6 (Ablex, Norwood, NJ) 71-80.

Carter, Anne P., 1989, Input-output recipes in an information economy, *Economic Systems Research* 1(1):27-43.

Castells, Manuel and Yuko Aoyama, 1994, Paths towards the informational society: employment structure in G-7 countries, 1920-90, *International Labour Review* 133(1):5-33.

Caves, Richard E. and Matthew B. Krepps, 1993, Fat: the displacement of nonproduction workers from U.S. manufacturing industries, in: Martin Neil Baily, Peter C. Reiss and Clifford Winston, eds, *Brookings Papers on Economic Activity, Microeconomics 2* (The Brookings Institution, Washington DC) 227-288.

Coe, David T. and Elhanan Helpman, 1993, International R&D spillovers, IMF Working Paper 93/84, Washington DC, November.

Cohen, Wesley M. and Daniel A. Levinthal, 1989, Innovation and learning: the two faces of R&D, *Economic Journal* 99:569-596.

Costello, Donna M., 1993, A cross-country, cross-industry comparison of productivity growth, *Journal of Political Economy* 101(2):207-222.

Davis, Lance E., 1986, Comment, in: Stanley Engerman and Robert Gallman, eds, *Long-Term Factors in American Economic Growth* (University of Chicago Press, Chicago) 149-161.

Economic Journal, 1992, Policy forum: the determinants of economic growth, 102:598-632.

Eggertsson, Thrainn, 1990, *Economic Behavior and Institutions* (Cambridge University Press, Cambridge, UK).

Engelbrecht, Hans-Jürgen, 1995a, The impact of international trade on the "in-house" transaction sector: evidence from the United States, 1985-1989, *Applied Economics Letters* 2:60-63.

Engelbrecht, Hans-Jürgen, 1995b, Corporate bureaucracies and United States Competitiveness, mimeo, School of Applied and International Economics, Massey University, July.

Engelbrecht, Hans-Jürgen, 1994, Private bureaucracies, organizational efficiency, and Australia's manufacturing trade structure: a comparison with Japan, *International Review of Applied Economics* 8(3):266-282.

Engelbrecht, Hans-Jürgen, 1992, New perspectives on intersectoral relationships between manufacturing and services, *Economics of Planning* 25:165-178.

Engelbrecht, Hans-Jürgen, 1991, Information services, private bureaucracies, and Japan's comparative advantage, *Review of Economics and Statistics* LXXIII(4):716-719.

Engelbrecht, Hans-Jürgen, 1990, A comparative analysis of the use of information inputs in the manufacturing sectors of Korea and Japan, *World Development* 18(1):77-89.

Engelbrecht, Hans-Jürgen, 1989/90, In-house information activities in an applied general equilibrium framework, *Information Economics and Policy* 4:305-323.

Engelbrecht, Hans-Jürgen, 1987/88, Information resources in U.S. manufacturing: a reassessment, *Information Society* 5(3):147-159.

Engelbrecht, Hans-Jürgen, 1986, The Japanese information economy: its quantification and analysis in a macroeconomic framework (with comparisons to the U.S.), *Information Economics and Policy* 2:277-306.

Fagerberg, Jan, 1994, Technology and international differences in growth rates, *Journal of Economic Literature* XXXII(3):1147-1175.

Fischer, Stanley, 1993, The role of macroeconomic factors in growth, *Journal of Monetary Economics* 32(3):485-512.

Freeman, Christopher and Luc Soete, 1990, Fast structural change and slow productivity change: some paradoxes in the economics of information technology, *Structural Change and Economic Dynamics* 1(2):225-242.

Furubotn, Eirik G. and Rudolf Richter, 1992, *The New Institutional Economics: A Collection of Articles from the Journal of Institutional and Theoretical Economics* (Texas A&M University Press, College Station).

Fuess, Scott M. Jr. and Hendrik van den Berg, 1992, The impact of transactions activities on U.S. productivity growth, *Economics Letters* 38:243-247.

Geroski, Paul A., 1994, Do spillovers undermine the incentive to innovate?, in: *1994 Conference of Industry Economics: Papers and Proceedings*, Bureau of Industry Economics/Australian National University.

Griliches, Zvi, 1994, Productivity, R&D, and the data constraint, *American Economic Review* 84(1):1-23.

Griliches, Zvi, 1992, The search for R&D spillovers, *Scandinavian Journal of Economics* 94(Supplement):29-47.

Gundlach, Erich, 1994, Accounting for the stock of human capital: selected evidence and potential implications, *Weltwirtschaftliches Archiv* 130(2):350-374.

Hansson, Pär and Magnus Henrekson, 1994, What makes a country socially capable of catching up?, *Weltwirtschaftliches Archiv* 130(4):760-783.

Hayes, Robert M. and T. Erickson, 1982, Added value as a function of purchases of information services, *Information Society* 1(4):307-338.

Helliwell, John F. and Alan Chung, 1991, Macroeconomic convergence: international transmission of growth and technical progress, in: Peter Hooper and J. David Richardson, eds, *International Economic Transactions: Issues in Measurement and Empirical Research*, chapter 11 (University of Chicago Press, Chicago and London) 388-436.

Imai, Ken-ichi, 1990, The information industry, in: Soesastro, Hadi and Mari Pangestu, eds, *Technological Challenge in the Asia-Pacific Economy*, chapter 5 (Allen & Unwin, Sydney) 74-89.

Irwin, Douglas A. and Peter J. Klenow, 1994, Learning-by-doing spillovers in the semiconductor industry, *Journal of Political Economy* 102(6):1200-1227.

Jones, Robert and Geoffrey Newman, 1995, Adaptive capital, information depreciation and Schumpeterian growth, *Economic Journal* 105:897-915.

Jonscher, C., 1983, Information resources and economic productivity, *Information Economics and Policy* 1:13-35.

Journal of Economic Perspectives, 1994, Symposia: new growth theory 8(1):3-72.

Journal of Monetary Economics, 1993, National policies and economic growth: a World Bank Conference, 32(3).

Jussawalla, Meheroo, Donald M. Lamberton and Neil D. Karunaratne, eds, 1988, *The Cost of Thinking: Information Economies of Ten Pacific Countries* (Ablex, Norwood, NJ).

Karunaratne, Neil D., 1991, A general equilibrium analysis of the Australian information economy, *Economic Systems Research* 3(2):171-185.

Kleinknecht, Alfred and Donald Bain, 1993, *New Concepts in Innovation Output Measurement* (Macmillan).

Krueger, Alan B., 1993, How computers have changed the wage structure: evidence from microdata, 1984-1989, *Quarterly Journal of Economics* CVIII:33-60.

Lamberton, Don, 1994, The information revolution in the Asian-Pacific region, *Asian-Pacific Economic Literature* 8(2):31-57.

Lichtenberg, Frank R., 1993, R&D investment and international productivity differences, in: Horst Siebert, ed., *Economic Growth in the World Economy, Symposium 1992*, Kiel Institute of World Economics (J.C.B. Mohr, Tübingen) 89-110.

Lucas, Robert E., 1993, Making a miracle, *Econometrica* 61(2):251-272.

Lucas, Robert E., 1988, On the mechanics of economic development, *Journal of Monetary Economics* 22(1):3-42.

Lundvall, Bengt-Ake, 1992, *National Systems of Innovation: Towards a Theory of innovation and Interactive Learning* (Pinter Publishers, London).

Machlup, Fritz, 1980, *Knowledge: Its Creation, Distribution, and Economic Significance, Vol. 1: Knowledge and Knowledge Production* (Princeton University Press, Princeton, NJ).

Machlup, Fritz, 1962, *The Production and Distribution of Knowledge in the United States* (Princeton University Press, Princeton, NJ).

Matthews, R.C.O., 1986, The economics of institutions and the sources of growth, *Economic Journal* 96(384):903-918.

McKelvey, Maureen, 1991, How do national systems of innovation differ?: a critical analysis of Porter, Freeman, Lundvall and Nelson, in: Geoffrey M. Hodgson and Ernesto Screpanti, eds, *Rethinking Economics: Markets, Technology and Economic Evolution*, chapter 8 (Edward Elgar, Aldershot) 117-137.

Miles, Ian, 1990, *Mapping and Measuring the Information Economy*, Library and Information Research Report 77, British Library Board.

Mody, Ashoka and Carl Dahlman, 1992, Performance and potential of information technology: an international perspective, *World Development* 20(12):1703-1719.

Mohnen, P., 1992, *The Relationship between R&D and Productivity Growth in Canada and Other Major Industrialised Countries* (Canada Communications Group, Ottawa).

Mokyr, Joel, 1992, Technological inertia in economic history, *Journal of Economic History* 52(2):325-338.

Nadiri, M., 1993, Innovations and technological spillovers, NBER Working Paper No. 4423 (National Bureau of Economic Research, Cambridge, MA).

Nelson, Richard, 1995, Recent evolutionary theorizing about economic change, *Journal of Economic Literature* XXXIII(1):48-90.

Nelson, Richard R. and S. Winter, 1982, *An Evolutionary Theory of Economic Change* (Belknap Press, Harvard University Press, Cambridge, MA).

North, Douglass C., 1993, The ultimate sources of economic growth, in Adam Szirmai, Bart Van Ark and Dirk Pilat, eds, *Explaining Economic Growth: Essays in Honour of Angus Maddison* (North-Holland, Amsterdam) 65-76.

Norton, Seth W., 1992, Transaction costs, telecommunications, and the microeconomics of macroeconomic growth, *Economic Development and Cultural Change* 41(1):175-196.

OECD, 1993, *Usage Indicators: A New Foundation for Information Technology Policies* (Paris).

OECD, 1986, *Trends in the Information Economy* (Paris).

OECD, 1981, *Information Activities, Electronics and Telecommunications Technologies, Vol. I: Impact on Employment, Growth and Trade* (Paris).

Oxford Review of Economic Policy, 1992, New approaches to economic growth, 8(4).

Parente, Stephen L. and Edward C. Prescott, 1994, Barriers to technology adoption and development, *Journal of Political Economy* 102(2):298-321.

Porat, Marc U. and Michael R. Rubin, 1977, *The Information Economy* (in Nine Volumes), U.S. Department of Commerce, Office of Telecommunications (Government Printing Office, Washington, DC).

Rabeau, Yves, 1990, *The Information Economy in Canada: An "Input-Output" Approach* (Department of Communications of Canada, Laval).

Radner, Roy, 1992, Hierarchy: the economics of managing, *Journal of Economic Literature* XXX(3):1382-1415.

Rebelo, Sergio, 1992, Growth in open economies, *Carnegie-Rochester Conference Series on Public Policy* 36:5-46.

Romer, Paul, 1994, New goods, old theory, and the welfare costs of trade restrictions, *Journal of Development Economics* 43(1):5-38.

Romer, Paul, 1993a, Idea gaps and object gaps in economic development, *Journal of Monetary Economics* 32(3):543-573.

Romer, Paul, 1993b, Implementing a national technology strategy with self-organizing industry investment boards, *Brookings Papers on Economic Activity, Microeconomics 2* (The Brookings Institution, Washington DC) 345-399.

Romer, Paul, 1986, Increasing returns and long-run growth, *Journal of Political Economy* 94(5):1002-1037.

Rubin, Michael R., 1988, The secondary information sector: its meaning, measurement, and importance, in: Meheroo Jussawalla, Donald M. Lamberton and Neil D. Karunaratne, eds, *The Cost of Thinking: Information Economies of Ten Pacific Countries*, chapter 9 (Ablex, Norwood, NJ) 101-110.

Schement, Jorge Reina, 1990, Porat, Bell, and the information society reconsidered: the growth of information work in the early twentieth century, *Information Processing & Management* 26(4):449-465.

Structural Change and Economic Dynamics, 1994, Special theme: institutions and economic change, 5(2):199-359.

Tomer, John F., 1993, A new rationale for industrial policy: developing the capabilities of the learning firm, *International Review of Applied Economics* 7(2):208-222.

Tomer, John F., 1987, *Organizational Capital: The Path to Higher Productivity and Well-Being* (Praeger, New York).

Wallis, John J. and Douglass C. North, 1986, Measuring the transaction sector in the American economy, 1870-1970, in: Stanley Engerman and Robert Gallman, eds, *Long-Term Factors in American Economic Growth*, chapter 3 (University of Chicago Press, Chicago) 95-148.

Warskett, G., 1984, The role of information activities in total Canadian manufacturing: separability and substitutability, *Applied Economics* 16(5):763-770.

Weder, Rolf and Herberg G. Grubel, 1993, The new growth theory and Coasean economics: institutions to capture externalities, *Weltwirtschaftliches Archiv* 129(3):488-513.

Williamson, Oliver E., 1993, The economic analysis of institutions and organisations — in general and with respect to country studies, OECD Economics Department Working Paper No. 133 (Paris).

World Development, 1989, Special issue: the role of institutions in economic Development, 17(9).

Young, Alwyn, 1993, Invention and bounded learning by doing, *Journal of Political Economy* 101(3):443-472.

CHAPTER 3

Telecommunications demand*#

James Alleman and Alain de Fontenay

1. Introduction

This chapter will review the session on demand and make recommendations on the direction of research based on Workshop commentators' views. All of the participants were experienced in telecommunications analysis; while not all were economists and only a few were practicing econometricians, all were well aware of the role demand analysis plays and all had requirements for its results. However, differences in our research background created, at times, small communication problems. Nevertheless, we rapidly discovered that we all had many interesting and relevant observations. One of the challenges we faced was to identify the contributions which various participants with varied backgrounds made and to find a common thread among them. While the process was somewhat tedious, the effort was worthwhile and the outcome fruitful. An achievement of the Workshop has been to create elements of dialogue between a "traditional" school which recognized its limitation but was able to produce some estimates and a more "prospective" school which would raise new questions too often forgotten.

Two concerns dominated the group's discussions on demand analysis. The first found its roots in the conventional approach to demand analysis which has been largely chronicled by Lester Taylor. (Taylor's recent book (1994) was used as the framework around which much of the Demand Session of the Workshop was organized.) The concern relates to what are the more appropriate extensions of the traditional demand models to address today's environment of rapid technological changes and the growing diversity in the

---

* This paper has benefited from discussion with Mark Schankerman and Paul Rappoport as well as the participants in the Telecommunications and Information Activities Workshop. A particular debt of gratitude is owed to Donald Lamberton, who organized the Workshop, encouraged this work, made many useful suggestions and kept us on schedule. Ms Barbara Miller offered excellent editorial assistance. The usual disclaimer applies.

This review draws from the transcript of the Workshop. The quotations are from the edited transcripts and are referenced by speaker.

services which are made available to end-users. Much of the focus centred around approaches which find their roots in the household production theory.

The second concern was characterized during the Workshop as "cultural". It is an approach which focuses on the underlying — and largely beyond economic constraints — reasons behind an end-user's use of telecommunica-ions and, upon reflection, it tends towards the attribute or characteristics approach. Where the econometrics concentrates on the demand for services in response to prices and income constraints, the convergence of the heretofore separate distribution media and the blurring of boundaries have led many analysts to look for new paradigms in studying and estimating the demand for telecommunications service. Those analysts recognize the turmoil in the telecommunications market: distribution media are changing (e-mail complements and substitutes in unforeseen ways for voice telephony and faxes), voice mail boxes (with their seemingly interminable tree of possible responses) are increasingly replacing people, CATV threatens over-the-air broadcast and in turn is threatened by digital broadcast satellites and broadband local loop capable of supporting telephone companies' video dial tone offerings.

The chapter ends with suggestions regarding possible directions for a research agenda in this area.

2. Historical perspective

2.1. Telecommunications environment

Twenty-five years ago, telecommunications services were virtually all of one type — voice communications. In addition, voice communications were provided to subscribers by a single operator, the subscriber's unique telephone company. The telecommunications industry has changed dramatically in the last few years. Furthermore the pace of change appears to be consistently accelerating. At the same time, the boundaries which used to separate clearly identified and well defined sectors such as telephony, broadcasting, CATV, computing and a variety of wireless technologies have become more and more blurred when they have not lost all meaning. As a result, market, services and legal definitions no longer obtain. Existing services are continuously integrated, merged and spun-off while new services are being constantly developed. At the same time, the market structure of the sector, the monopoly PTT has given way to an increasingly complex world characterized by the daily diet of new alliances, mergers and joint ventures. This global convergence, at the level of both service provider and of services available to the end-user, is well documented by many on the telecommunications scene.[1]

2.2. *Demand analysis: implications*

The implications for demand and forecasting had been even more drastic. Where the analysis of the past, while typically making a sophisticated use of large and often relatively reliable databases, was nevertheless fundamentally simple, today's analysis has to cope with questions which are orders of magnitude more complex while benefiting from poor if not non existent databases. In addition, new products and services confound analysis since they may be close substitutes for the existing services or require the existing services for their production.

Since the PTT was the monopoly, the sole provider, the analyst did not have to concern herself with the question of market share. Firm-level elasticities were the market's elasticities. Today, the analyst must concern herself with inter-firm cross-elasticities.

Pricing was a relatively simple task but for the progressive introduction of certain new pricing schemes such as WATS. Prices were mostly linear and the diversity of price structure was largely amenable to conventional price index analysis.

Product differentiation played at most a minor role, a voice call was a voice call. The former homogeneity of the sector's output is giving way, through the impact of today's competition, to a range of partially substitutable service offerings created by operators who attempt to differentiate their services from those of their competitors.

Market segmentation was effectively limited to a differentiation between business and residential customers. Aggregate forecasts were "good enough", market segmentation and specific elasticities and forecasts were not required. Revenue requirement, an expression which has become quaint in an increasingly deregulated world, was the primary motivation for the production of market forecasts. The impact of forecasting errors was limited and the operator was essentially able to revise its "revenue requirement" and use its monopoly power to achieve the desired results. The regulatory process, in the few places where there was one, was straightforward. It focused primarily on a set of prices which were seen as particularly sensitive from a public policy perspective. Moreover, the regulatory process did not allow for price variations, price discrimination or significant non-linear pricing.

How things have changed! Niche markets are everywhere. Interconnection, at the "correct" price, is demanded by new operators to provide their customers access to subscribers of the dominant operator. Conventional economies of scale and scope, the object of in-depth analyses prior to AT&T's divestiture, are not anymore the object of the kind of interest it had over a decade ago. Whether those economies have disappeared or, at least, lost much of their significance, whether the pace of technological change is now dwarfing them through dynamic efficiency gains, or whether the competition by forcing the operators to meet growing end-user demand through product differentiation, stressing new dimensions of the market processes, the focus of the analysts and of the market has been away from those economies.

On the other hand, conventional demand measures such as price elasticities and cross-elasticities are much more critically important than in the past even though we have already seen that they have become much harder to estimate where they can be defined! (For example how does one define the demand for Reach Out America, Friends and Family, etc. services offered by the interexchange carriers in the United States?)

2.2. Taylor's survey and identification of problems

Taylor's (1994) recent book was reviewed and elegantly summarized by Lamberton with a view to better understanding the current state of telecommunications demand analysis and identify today's most significant problem areas. Since the publication of Taylor's first book on telecommunications demand analysis in 1980, the sector has changed dramatically and dynamically. Traditional telecommunications demand analysis was undertaken primarily for regulatory purpose, as a means to identify revenue requirements. The models were developed within the context of a monopoly and the results were typically used to justify regulatory policies, usually rates. Results were used to a much lesser extent for business, marketing and policy purposes.

The regulatory community was primarily concerned with the rate level. Once competition began to grow and to be seen by incumbents and, sometimes regulators, as a threat to the regulated markets, greater attention — especially on the part of the dominant operators — was given to the rate structure — after all this was one of the major ways new providers had been able to make inroads. It was then that demand analysis flourished in the industry.

Taylor (1980) included virtually all the telecommunications demand papers in his original survey; in the revised edition (1994), he had to narrow his focus because the literature was so vast as he noted in his introduction. Taylor identifies what he sees as two major demand developments since his first book. The first is a much greater depth in today's knowledge of demand for residential access. For him, the second is what he considers to be a significant start in quantifying the structure of business demand. Research has confirmed and sharpened previous estimates of the price elasticities for medium to long-haul toll usage. Today's literature has made substantial headway in the short-haul long distance. In addition, it now provides a range of estimates of cross-price elasticities between measured and flat rate local service; local service and toll; local use and toll, and key, PBX and Centrex systems.

There are already a few studies of the demand for custom calling features (Kridel and Taylor, 1993), where the features are sometimes considered independently and sometimes in a broader context of interdependence. Unfortunately, these changes in the industry structure which have led to the creation of new services at an unprecedented rate, even by comparison with the beginning of the century, have also led to a much higher level of restriction on the access to the data, today's most significant threat to research on demand for telecommunications services.[2]

While Taylor points to what he sees as a fair measure of success in modelling, the key phrase he uses is "... has made a start". Econometrics methods have also advanced during this period, as they must to be able to handle the new requirements for analyses. Logit and probit models have been extended to address questions such as the demand for intraLATA toll, the choice of interLATA carrier, and the effect of exchange and interexchange services being integrated in Japan (Train, 1995a and b).[3]

The problem areas identified by Taylor (1994) were: (i) the demand for business services, (ii) the separability[4] between telecommunications and other goods and services in a household's budget, (iii) the dynamics of information exchange in the demand for telecommunications services where usage stimulates its own usage — Lamberton reminded us of that dimension when he cited Taylor, "calls give rise to further calls, quite independently of price and income" — (iv) option demand, and (v) service provider-specific demand elasticities.

Marc Ivaldi discussed the use of socio-demographic variables to predict usage of residence telecommunications demand. He talked about variables traditionally used in telecommunications demand models, variables which are easily observed and measured, like age or the number of children. He went over their use in a model he worked on in France, pointing out that they were not reliable in many situations and could not be employed to enhance the predictability of the usage. He felt that the data to build the desired subscriber-level demand model are available, but that they usually imply complicated and hard to manage models. He had explored introducing iterations into the model to bypass the problem. The model he had formulated was essentially a model with incomplete information.

In the past analysts had a monopoly, i.e., a stable market structure. In addition there was only a small set of commodities which was easy to model. Today's challenge is that this structure no longer obtains. Thus, Taylor's book is more an analytical history — almost the testament of a dying art in the opinion of one of the commentators. Today's technologies have the potential to provide a much wider range of services to the extent that what are "other" commodities might not be well defined anymore.

3. Rationale

Why do we do demand analyses at all? Four rationales come to mind:

3.1. Business rationale

In order to effectively market services, set their prices and obtain regulatory "buy-off", the demand characteristics of business customers are critical. There is a need to know how business and other customers react to a change in price and/or to a change in their economic fortunes, i.e., the firm needs to have a good sense of what the price and income elasticities of demand might be. In fact the service provider needs to know much more. For instance, a good understanding of the market would include the identification of the subscribers who will be affected by price realignments. All of these factors, together, contribute to estimating the revenues new services might generate. To that end, furthermore, the operator will have to find out whether and, if so, to what extent the service might be a substitute which may compete with existing service offerings — this kind of consideration is seen as going far in explaining the slowness with which N-ISDN has been deployed for instance. These factors continue to be the main strands of analysis in the industries. In and of themselves, they would have gone a long way in explaining the difficulties faced by demand analysis today. However, they constitute only one part of the issue. With the erosion of the monopoly and the growth of competition and with the transformation of the regulatory process from the regulation of a firm — the monopoly — to the regulation of the industry and of the competition in its midst, most of the databases used for demand analysis and the analyses themselves have been "privatized" with few discussions in public forums of their results and even of the methodology used. Indeed, the market structure has an important impact on demand. The firms undertaking the analyses do not wish to reveal the results to their competitors, preferring to make decisions with significant unknowns rather than collaborate and provide access to sources of information of use to its competitors.

3.2. *Policy rationale*

The policy rationale parallels the business rationale above and the academic rationale below, with a major exception. The goal of policy is to help us determine where we want to be, the market structure we would like, the regulatory method, if any, which would best help us and the prices to be charged. The previously accepted notion that the traditional demand (and cost) analysis can achieve this end appears to be in doubt. Demand is increasingly seen as endogenous. The same is also true for the technology, hence the cost functions. Together with an understanding of the market structure, they may help better understand "where we are". A much broader analysis would be required to begin grasping both "where we should be" and how to reach it.

The traditional view that demand and cost estimates could be used to determine desirable policy, for example the optimal rate structure, collapses in the kind of environment which has evolved in recent years and it is not possible anymore, using traditional methods, to help the policy-maker who wants to know were we should be. Nevertheless, while the analyst might have to be more humble as to what she can produce, her contribution, in view of the complexity of the environment within which we all evolve, is probably far more significant than in the past.

3.3. *Academic rationale*

To understand the complexity of the demand for telecommunications services one must strive to understand human behaviour — cultural and psycho-social as an integral part of the general desire in understanding communication behaviour. A concern with the behaviour dimension of economic agents *vis-à-vis* the information is part of this research and we focus here primarily on the concerns of the non-economist academics who have been more interested in the demand for information.

In the case of the business rationale, as we saw above, clear objectives can be and are specified even if they might not always be reached. The academic understanding of demand can be quite diffuse. It cannot be circumscribed to one discipline or one methodology. It considers questions as varied as to whether women and men use the phone differently, the kinds of communications which are transmitted, their information content, and whether people-to-machine communications are alienating. Addressing those questions is not purely a free exercise and the results can be used in the business context as well.

3.4. Technical rationale

How are the communications needs to be met? Are substitute technologies available? Can the distribution of information be conveyed differently? What can be done with the technologies? What new services are offered? What service can be provided more efficiently? The technical rationale overlaps substantially with the business rationale and, to a lesser extent, the academic rationale. Nevertheless we chose to identify it as a separate element to reflect the fact that technologists may — and have to a great extent in the past — develop services with little or no reference to business consideration or to observed consumer behaviour.

One can think of N-ISDN as a technology which is still trying to be fitted to consumer demand. However, the most glaring example is the introduction of the "picture phone" by the Bell System in the early sixties, and more recently, as Sam Weerahandi *et al.* (1994) report, in the Bellcore Laboratories. (These researchers are concerned with the linkages between the boxes and the means to provide access to the information.) The means to store, distribute and retrieve information have completely changed in the last few years. These changes have been shifting dramatically the quantity, quality and cost parameters faced by operators forcing them to rethink their strategy in fundamentally new ways just to survive and succeed.

Information is a unique commodity, with all its cost or almost all of it being carried in the first unit. It creates extreme situations in which the operator, at times, is able to extract large rent gaining at the same time large market power to the detriment of society and its consumers. This is the kind of situation Michel Menou introduced in the discussions, providing examples drawn from his professional life in less developed countries.

4. Household production theory

4.1. Household production theory: concept

While the conventional approach is based on Marshallian demand, there is an alternative approach, with its roots in the household production theory, which may be particularly useful to model demand in today's competitive and dynamic telecommunications market place.[5] Rod Maddock pointed out the fundamental weakness with the narrow application of the Marshallian demand relative to the household production theory by reminding the participants of the "... distinction between a Marshallian and the Lancaster (1971)[6] utility function ... if we keep adding products into the [Marshallian] utility function, the happiness of the human beings increases just because of the addition of new services ... [T]he utility function is changed simply by the addition of new products. Consequently, the overall effect of possible choices is increased together with the happiness of the agent and consequently cross-elasticities are changed as well."

Maddock went on to stress the unique problem in the information sector, observing that the "problem ... is again the dynamic notion, of rival products which provide rival services and we should try and understand how the dynamics of the behaviour of the individual is affected by the changes in the utility function for information, and the changes in utility function for information services, which I believe is terribly complicated."

4.2. The characteristics/attributes approach: an application

Michel Menou presented a different view of demand analysis. He sees telecommunications as a means to an end. In his view, the analysis of demand must proceed from the knowledge of the underlying utility of the communications. A view that reflected the characteristics of information and how it should be utilized, this is in contrast to the traditional view of demand. Only by understanding the demand and infrastructure for information (information being defined as that combination of technological means of carriers and contents, according to Menou), and, perhaps, communication including telecommunications, can one understand demand. Unless one knows the people as individuals, groups, organizational structures, how can one determine what is occurring. "The first basic component is the information itself. What goes into the pipes ..." while foreign to the traditional view of demand analysis, it has its antecedents in the characteristics or attributes approach.

Sam Weerahandi is one of a number of people who is pursuing the approach of modelling the characteristics or attributes demanded by end-users. His concern, however, originates from the characteristics of technologies rather than those of services, product or information.

In their empirical work, Weerahandi *et al.* (1994) used a model which bears much in common with the household production approach to identify the demand for telecommunications services. They begin by investigating the functionalities or characteristics which are arguments of end-users' utility functions. Then they consider the various ways those functionalities or characteristics which contribute to the end-user's satisfaction can be produced from services available on the market and across technologies.

He used his own Bellcore (Bell Communications Research) experience to illustrate one of the uses of his approach. He could observe first hand the problem created with "bottom up" forecasts where typical market projections for new services are made without accounting properly for cross effects due to other services which, while being different in some sense, are nevertheless functionally equivalent but analysed from the demand perspective independently. For example, the applications which are expected to drive the SMDS market, e.g., video conferencing, imaging, etc. were evaluated and quantified. At the same time, forecasts were being produced for services such as DS1 and DS3, using essentially the same source of demand.

Weerahandi's insight is to characterize each of the competing services as a unit — that is all of the services are treated as being provided by the same supplier. Services are characterized one by one in terms of their attributes, e.g. dedicated or switched where DS1 is dedicated while SMDS is switched. The customer only pays for the portion that she demands as in the case of a particular bandwidth. Depending on the application, one customer may require 1.5 megabits of bandwidth while another requires twice that amount. This approach together with a sub-optimal programming technique made it possible to estimate the fraction of the demand which will go to SMDS, the fraction which will go to dedicated services like DS3, DS1, etc.

His attribute model enabled him to estimate the market share for each different but competing service. While this has been applied in other contexts before, it is extremely well adapted to today's telecommunications.

Weerahandi noted that "It was a matter of dividing the pie." However, they have gone only part of the way: "What was missing in that study was actually estimating the total demand to cover all of these applications ... When we added the new services which provided high bandwidth services there might be some additional demand. There might be some new applications that will occur as a result of availability of the new services ... [however] this was not addressed in the paper."

The participants went on to consider possible extensions of Weerahandi's work. Marc Ivaldi, placing the discussion back into the Marshallian context, proposed "... that as soon as we assume that a firm or individual is using the different type of technology as input to build a certain level of information, then the problem of using different attributes [of the technology] is not so important ... it is more a problem of building up a demand system, and a problem of measuring the different quality of communication of information rather than just describing the technology."

4.3. Household production theory: some further thoughts

Michel Menou presented what seemed at first a different view of demand by stressing a number of characteristics most economists fail to consider in demand analysis. Nevertheless, it was particularly interesting that many of his concerns as well as many of the concerns of those who approach the subject from a more historical or sociological point of view did seem to point in the same direction as some of the recent conventional research. In Menou's view, the analysis of demand must proceed from the knowledge of the underlying utility of the communications which is essentially what the household production theory strives to achieve. While the concept is the same in both the work which has been building upon the theory of household production, and Menou's approach, which focuses on the transformation of the goods and services purchased by the consumers to derive the desired satisfaction, there is a fundamental difference between the two approaches The strength of the household production theory in terms of supporting econometric applications is its fundamental but unstated assumption that only the attributes of a telecommunications service affect the consumer's satisfaction, i.e., that the technology of the household production is costless and transparent to the consumer's decision process.

For Menou, on the other hand, the technological means through which the information is transported and processed cannot be separated from the content itself. Together they form what Menou defines as "information". The implication of his hypothesis is to reject both the conventional Marshallian demand models for focusing exclusively upon the commodity the consumer purchases on the market and the household production theory for considering exclusively the attributes those commodities can produce as a result of the household production process. Menou would need both the Marshallian characterization and the household production characterization at the same time to arrive at an empirical test of his fundamental hypothesis. At this stage, however, the need to consider the household production process is still only a conjecture.

Menou used examples to challenge the universal concept of abstract demand, whatever technology, used in the conventional approaches. He noted that anyone who has used a telephone in Cairo or Rio de Janeiro, as opposed to the United States or France, knows that the service they consume is not exactly the same service nor can one utilize it in the same manner. In another example, he used France and the access to train information: "To make things a bit more concrete, ten years ago in France when you wanted to know what time a train left, you would call the station and a nice lady would answer that the 8:40 a.m. train is actually leaving at 8:40. Now you call the station and you get a tape that says, 'Sorry we are very busy, would you please call 3615 to get the information.' 3615 is the Minitel journal service for this type of information, so you get into the Minitel after having rolled through a number of screens, you discover that there is no train ... There are many other examples of that kind", Menou added. While this is correct, there are a number of alternative ways to explain the phenomena. First, the French railway system is government-owned and, facing no substantial competition, it has not had to be customer-oriented. It may just as well be that customers are not willing anymore to pay for information provided by live operators. It may also be that for most customers most of the time, Minitel offers a superior alternative: one of the commentator's childhood memories is that the "nice lady" would take her time to answer if she answered at all, that she was likely to be particularly impolite and unhelpful while complaining about everything; the Minitel, at least, is functional.

5. Demand analysis: new perspectives

5.1. Introduction

The Workshop went at times beyond the more conventional models to raise more fundamental questions, questions which may not lead, immediately at least, to new econometric models but which, through time, contribute to a greater understanding of the problems at hand. Issues which it identified included the interdependence between the evolution of demand and the evolution of technology and whether the utility function from which demand is derived is time-independent or path dependent. It is interesting to note that it is to a great extent the third of Taylor's areas of concern, the dynamics of information exchange in the demand for telecommunications services where usage stimulates its own usage which primed the participants to go beyond Taylor's considerations and focus for a time on other economic and "cultural" areas.

While some wondered what are the permanent elements in people which demand communications — both accessing information and exchanging information, on the ground that, in telecommunications the greatest stability could be found with the end-user, an observation which received some support from Oniki and Cracknell, Antonelli questioned the notion of "permanence" in demand.

Linking the market structure of competition and technologies in the analysis is the type of demand analysis required in the future (Swann, chapter 5; Nightingale, chapter 6). While we are far from that in our empirical work, the matter was the object of substantial discussions. This linking would be complex, requiring sociological work on why people behave the way they do, and in particular ways — why do they demand and use a telephone or more generally telecommunications.

For instance, one of the commentators noted during the discussions that "... [P]eople demand mobility. People are able to use the phone to do more things. However, we do not understand precisely in what way ..." People are willing to pay substantial sums for communications mobility, for example, but we have little understanding of this.

5.2. Interdependence of demand and technology: long run implications

Demand is fundamentally studied "given technology". A new technology may affect the cost of producing some service or still make it possible to combine services in new ways. However, that technology requires demand to render it viable. This issue was the source of some discussion. One commentator pointed out that the dichotomous classification of technology and demand is inadequate: "Demand is affected by technology and vice versa." In all the models we have considered up to this point, including Weerahandi's attribute model, take the technology as given and derive demand, yet, especially with new technologies, it is that interaction between technology and demand which is critical. As analysts, we cannot escape considering the dynamic problem (see Swann, chapter 5).

Two dominant issues were identified. First, as "... the general heading would be product development, ... innovative product development in the area would of course affect demand and that is perhaps the most interesting aspect of it." Then "the other thing ... especially if you look at the demand for telecommunications services in the industry, has to do with the level of knowledge, the knowledge that is used in production." Production has to be organized, placed into a framework with classification so that "the receiver has to understand — essentially ... a language problem."

One analogy would be printing technology. Once Gutenberg began printing, the demand for printing grew and spread. The diffusion process was essentially a function of the ability of people to read and to write. We can observe a process very similar today "in industries, of course" where "a lot of production processes are being put on code, that were earlier done in a different way." What is relevant here is that such a step, in turn, increases the ability to communicate information on how to run the process, and how to transfer that knowledge, etc. resulting in further growth for telecommunications services. The expectation is that it has the potential to be an important factor in the future. This dimension of demand and its implication for the factors which would impact communication and telecommunications services is generally not examined.

5.3. The stability of communications expenditures through time

Whether or not individuals have a static need for communications led to the examination of the dynamic nature of the demand problem and its complexity. Oniki introduced the issue, using the Japanese experience to illustrate his point. First he noted the extraordinary stability of the fraction of income Japanese households have spent on communications over the past 40 years, i.e., from the 1950s to today. Thus in the 1950s telephone communication, broadcasting, and ordinary mail constituted approximately one and one-half per cent (1.5%) of the consumer's income. Today it is slightly above two per cent (2%) which is extraordinary considering the change in Japan's real income in the last 40 years — five or six times the magnitude in real terms.

To provide a better perspective, Oniki noted that "on the other hand, the amount of money spent on transportation, moving around to meet other people to exchange information, has increased remarkably. Forty years ago it was about two per cent (2%) ... [T]his was the time when automobile transportation was almost nil. Today people are spending about twelve per cent (12%) of their income for transportation alone, including automobile transportation and public transportation."

In the same period, he pointed out that while "... the time people spent ... [travelling], excluding business activities, but including commuting and pleasure," for "... the average person, including babies and old people ..." was a half hour per day, today, it has reached a whole hour. That means that, in contrast to communications, "the time spent on transportation has doubled in the past 40 years." Oniki's objective was not so much to draw any definite conclusions from these statistics, but rather to stress the stability of the communications series and the instability of the transportation one. It was to invite analysts, once again, to take a longer term perspective when analysing a process as complex as demand.

Oniki had, however, a complementary objective, a policy objective which he illustrated with Japan. He noted that if Japan diverted one per cent (1%) of its transportation budget to communication, it could easily finance a nationwide fibre optic network.

Cracknell commented that the issue Oniki was addressing is unlikely to be unique to Japan and the statistics cited by Oniki might hold across countries, e.g., in the UK not only are telecommunications expenditures a stable percentage of income but that percentage is generally comparable with those of Japan and the USA.

5.4. Utility function: time-independence vs path dependency

For Antonelli the question raised by Oniki and Cracknell's observations had to do with the nature of the consumer's utility function through time. Demand analysts typically assume implicitly that a utility function is constant, unchanging over time. Thus, we have a long term definition of the needs of human beings and then we simply want to fit the supply into those potential demand curves. Instead, one may want to assume that it is dynamic, evolving through time.

Antonelli proposed three reasons why that may be a superior way to study demand. First, common observation indicates that firms are willing to spend resources to induce changes in the utility function of consumers. He illustrated his first argument by suggesting that advertising is nothing else than a way to modify such utility functions. Second, the utility function of a particular consumer is directly affected by the interdependence of that customer with other customers. The classic example of that kind of interdependence is the desire to "keep up with the Jones". Certainly, advertisers often use that strategy to motivate consumers to buy the product they advertise. A purely static utility function would have to be based on the premise that the utility functions of human beings are strictly individualistic. Thirdly, experience and learning are elements which transform utility functions through time; human beings learn to appreciate new goods and new services and, consequently, the satisfaction they derive from them transform their utility function.

For Antonelli, Oniki's data should lead to the following question: Should one assume that the Japanese utility function had imbedded in it, already such a large portion of transportation services in 1950 or should one assume, rather, that the Japanese had learned through inducement and/or interdependence with others to consume more transportation services?

5.5. Utility function: cultural and gender dependencies

As telecommunications penetration is growing everywhere and as the differences observed by Menou between Cairo and Rio de Janeiro on the one hand and Paris and New York on the other hand are beginning to become less significant, a question lingers in the mind of the analyst, namely whether, if demand can properly be modelled in one place, the results can be used in another setting, say a different country.

Ann Moyal placed the question in a complementary but comparable context when she reported on her research on the difference between how men and women "consume" telephony. She noted that "it ... [raises] a wide question ... studies done on women and the telephone in Australia and ...

feedback from other countries" indicate that the two genders use the telephone differently. "Women use the telephone and all communications services quite differently from men and they certainly do not use it just for that sort of instant information, but for a whole set of nurturing and caring responses and the whole business of not travelling but human communication." She also noted that, while demand analysts appear to have failed to look at such factors, firms which are competing for their market, including telephone companies, are increasingly making use of these results in their marketing.[7]

Gender difference is certainly a factor which will become increasingly significant with the increasingly diversified telecommunications good in an environment which is ever more competitive. It is not clear, however, that it can help us understand whether culture matters in demand analysis.

For researchers like Moyal, Robinson, and Menou, it goes without question that "culture" matters. Thus, Menou noted that "one dimension which is very often missing, systematically missing, is the cultural dimension ... a cross cultural perspective." For others like Maddock, the question is just as clear: "it is easy to observe that 'cultures' are different, and therefore we are going to behave differently ... The challenge is to find the commonalties across cultures. But the answer is not as obvious ... People acknowledge that the demand for telecommunications is a derived demand. The real question is what is it derived from? This is where the cultural dimension must be placed ... And just how sharply one can draw a boundary ... [This] is a major issue."

6. Demand analysis: other issues

6.1. Introduction

As we said earlier, the Workshop went beyond the more conventional models and raised more fundamental questions. The answers to those questions may not lead, immediately at least, to new econometric models.[8]

6.2. Micro-studies: telecommunications and R&D

Workshop discussion proposed the following comparison between the transportation of physical components in manufacturing and R&D: "We readily accept that you transport components, ... when you assemble a car, for example, to use the advantage of specialization. But much of this can also be seen in information-based services ... [A] typical example is software development and the increasing software product content, even in hardware product offerings." This inspiration came from some recent micro-studies on business demand for qualified knowledge. Those studies led to efforts to identify better the extent to which telecommunications is used for research and development. R&D is a personal intensive type of activity which appears particularly dependent on physical contact yet the telecommunications network is a vital link for the transportation of quantifiability of knowledge.

Returning to a software example, it was noted that, from a R&D statistician's point of view, the majority of the [non-voice] telecommunications demand is simply shipping around modularized pieces of software ..." But with data communications — but excluding voice telephony — that is the most important usage of telecommunications services which can involve a huge volume of data — as in the case of debugging where one sends out large quantities of data for testing, even outside the organization.

People are worried about cost, it was argued. It was the cost of using CPU time at distant locations, cost which tends to increase dramatically with the volume of data rather than the cost of telecommunications that was of concern. This analysis was based on comments R&D managers had made. The fundamental transformation we are going through is the fact that, while, in the past, information was primarily sent, today it is increasingly screened or searched. In this context, the telecommunications system has become an extremely sophisticated transportation system. A dramatic increase is occurring in the search for information, an operation which tends to be inexpensive in telecommunications terms but rather costly in processing terms.

The bottom line of this proposal was that, in the context of R&D at least, transport, plain transport, rather than interaction is the key demand on telecommunications.

6.3. Universal service

With the introduction of competition into the telecommunications industry, much of the concern over the last two decades and still today is with "universal service" or high market penetration for telephone service.[9] The implication is that it would benefit the poor, hence the large amount of support it has received from consumer advocacy groups. At the time of divestiture in the US, subscriber access and the elasticities of demand for access were the object of major studies. Perl (1978 and 1984) initiated demographic estimates of penetration. More recently, Hausman *et al.* (1993) have found cross-elasticity effects of toll rates on universal service.

Little is known and much needs to be done to understand the relationship. A recent survey by Mueller and Schement (1995) questions whether low income groups have the kind of demand for telephone services consistent with universal service provision — their conclusion is that, for many low income families, such callers as bill collectors and tele-marketers together with the lack of control over actual expenditures and the telephone bill is leading low income consumers to prefer cable television to telephone services.

6.3. Interconnection

Public utilities of the past were sole and complete providers, the only point of supply for the end-user. Nowadays, while the former monopolies are still everywhere the major players, things have changed with layers upon layers of competing service providers such as resellers, competing network providers, and other service providers. Switching and transmission have become commodity markets of their own right.

Incentive regulation is the mechanism of choice for regulatory control of the "dominant" firm and even the latter has been given increasing freedom to price the services it deploys as it sees fit. As a result, today's retail prices are controlled by operators, not their regulators. Previously non-existent markets now exist to serve niche and specialized needs. Competition in and of itself has created new services such as the redundancy to have an alternative carrier for security purpose.

Price realignment is helped by phenomenal increases in efficiency. Prices are critical at all levels. Wholesale markets have been created which may even provide service to the former monopoly. It furnishes new service providers which are striving to enter specialized and niche markets with an alternative source of supply.

In today's world of networks of networks and systems of systems, the key pricing issue has become the pricing of interconnection, at first primarily for access but increasingly to achieve the transparent communications consumers demand. Those access and interconnection prices are seen as being at the heart of competition and of the evolving market structure. They are central to the control of predation, optimal entry and economic efficiency. In most countries where incentive regulation has been adopted, they are the only prices which remain under the scrutiny of the regulatory authority. This is true even in a country such as New Zealand which has achieved deregulation by doing away with industry-specific regulatory bodies and which is now actively reviewing alternative approaches to interconnection pricing. Any determination of these prices which utilizes the result of recent work in economic analysis is essentially highly information intensive, requiring detailed demand and cost data and often cross-elasticities.

7. Conclusions and recommendations

7.1. Introduction

The Workshop made it possible to gain a better insight as to the elements which a framework for demand analysis should consider, even require, to address today's telecommunications environment. This outcome was primarily the result of two factors. Taylor's 1994 book on demand analysis provided a constructive structure around which to organize discussions. Then the diversity in the participants' backgrounds led to exploration of questions which are not generally considered.

To wit, the coherence we are presenting in this paper was the result of our work after the Workshop. We do not believe that the participants, whether they belong to the group of those who are at ease with conventional quantitative demand analysis or to the group of those who approach demand from its historical and sociological perspective, came out of the Workshop with the feeling that there was a high level of coherence in the views expressed.

Even if sociologists and historians may still have some difficulty in understanding the discipline a market imposes, they should be able to see many of their concerns being increasingly tested and, as appropriate, incorporated in the demand analysis.

While we talk of expanding the set of hypotheses to evaluate in the course of demand analysis, we must remain aware that, at the same time, the estimation problem has grown in complexity. Access to data is restricted and the continuous introduction of new services and new pricing plans limit what can be achieved with time series data. More significantly, the definition of the "economic good" which is being traded in response to new pricing plans is becoming more and more complex.

In spite of these new problems, we were able to identify new areas for research beyond those which had been named by Taylor (1994).

7.2. Household production methodology

There is little doubt that, as we move to highly disaggregated cross-sectional data, the way to address the service definition and pricing problems will increasingly be through models derived from the general household production approach. Those models appear particularly well designed for an environment where there is a web of services, most varying from the others by relatively small features.

7.3. *Endogeneity of the household production*

Many of the concerns expressed by non-economists in the context of demand modelling can be rephrased using the household production structure. This approach, however, addresses only part of the problem. The non-economists' vision tends to question the implicit assumption of separability between the consumer's utility function expressed in terms of attributes and the household production model expressed in terms of traded goods and services. For them, it is the wider context of this process which must be understood, if one is to hope to model demand. Their insight may become particularly significant as we move toward new services which are not well understood as in the case of Internet.

7.4. *Endogeneity of demand*

Neoclassical economics consider consumers whose tastes are invariant to the environment. In a world of rapid technological change and evolving boundaries, to consider taste invariant — even at the level of the attributes — is an assumption which appears to be hard to defend. In fact, demand and its formation is a direct input in the technological path as can be observed with the advance gained by GSM over TDMA and CDMA for instance.

7.5. *Path-Dependency*

The issue of path dependency does not arise except at the level of technology through which services are offered — we mentioned earlier the advance taken in the last three years by GSM, even at the research level, relative to CDMA and TDMA but also at the level of demand — but also at the level of consumer tastes. Arguments can be developed which present the consumer utility function as path-dependent upon technological solutions.

Notes

1. It is not our purpose to document the dynamics here as they are well known to those in the industry, but to point out how they affect demand estimation.
2. In a review of estimation methods, the majority of the studies reviewed were proprietary. The reviewer had permission to state the results in a qualitative way, but no public document with the actual data or coefficients exist. (Train, 1995a and personal correspondence).
3. In addition to the previously logit and probit analysis methods, Train (1995a) also cites probit simulation and the development of better simulation tools.
4. Recently, Wolak (1994) tested explicitly for separability and rejected the hypothesis, although one author does not believe the data set utilized was sufficiently robust to support the conclusion. With the methodology of the above cited paper applied to a richer data set significant light could be shed on this topic.

5. While the participants would like to see cross-culture databases, only recently have databases been available across telecommunications services. In the United States, PNR & Associates has recently developed a large database (32,000 observations) across all services, both cable and broadcast TV, mobile, telephone and other telecommunications services. This is the only database of this type of which the authors are aware.
6. As Donald Lamberton indicated Australian economist, Duncan Ironmonger anticipated the Lancaster characteristics work, his ideas being formulated at Cambridge UK during the period 1957 to 1960 and incorporated in a doctoral dissertation submitted in 1961 — see Supplementary References.
7. This is not unique to telephone conversations as Deborah Tannen's (1992) research has shown.
8. See Train (1994) for a review of the latest econometric models being applied in the telecommunications context.
9. One author (see Alleman, 1995) does not believe the goal of Universal Service is well conceived but must be taken seriously because most policy-makers are not convinced of its inappropriateness.

References

Alleman, James H., 1995, Interconnection & incentive regulation: an overview, International Telecommunications Society Workshop, Wellington, New Zealand, April 10-12.

Baumol, William J. and J. Gregory Sidak, 1994, *Toward Competition In Local Telephony*, AEI Studies in Telecommunications Deregulation (MIT Press, Cambridge and London; American Enterprise Institute for Public Policy Research; Washington, DC).

Duncan, Gregory M. and Donald M. Perry, 1994, IntraLATA toll demand modeling: a dynamic analysis of revenue and usage data, *Information Economics and Policy* 6(2)(July):163-178.

Gasmi, Farid and William W. Sharkey, 1995, Toward the endogenization of firm's costs in empirical studies of technology, International Telecommunications Society Workshop, Wellington, New Zealand, April 10-12.

Hausman, J.A., Timothy J. Tardiff and Alexander Belinfante, 1993, Effect of the break-up of AT&T on telephone penetration in the U.S., *American Economic Review* 83(May):175-184.

Kridel, Donald J. and Lester D. Taylor, 1993, The demand for commodity packages: the case of telephone custom calling features, *Review of Economics and Statistics* 75(2) (May):362-368.

Mueller, Milton and Jorge Regina Schement, 1995, Universal service from the bottom up: a profile of telecommunications access in Camden New Jersey, World Wide Web: http://ba.com/nr/95/feb/rutgers.html.

Perl, L.J., 1978, Economic and demographic determinates of residential demand for basic telephone service, National Economic Research Associates, Inc., March 28.

Perl, L.J., 1984, Revisions to NERA's residential demand for telephone service 1983, prepared for the Central Services Organization, Inc. of BOCs, National Economic Research Associates, Inc. White Plains, NY. December.

Tannen, Deborah, 1992, *You Just Don't Understand: Women and Men in Conversation* (William Morrow, Ballantine, New York).

Tardiff, Timothy J., 1995, Effects of pre-subscription and other attributes on long distance carrier choice, *Information Economics and Policy* 7(4):353-366.

Taylor, Lester D., 1980, *Telecommunications Demand in Theory and Practice* (Ballinger Publishing Company, Cambridge, MA).

Taylor, Lester D., 1994, *Telecommunications Demand in Theory and Practice* (Kluwer Academic Publishers, Dordrecht/Boston/London).

Train, Kenneth, 1995a, Econometrics for a competitive age, *Proceeding of Forecasting and Demand Analysis in the Global Telecommunications Environment*, International Communications Forecasting Conference, Toronto, June 13-16.

Train, Kenneth, 1995b, Simulation methods for probit and related models based on convenient error partitioning, Working Paper 95-237, Department of Economics, University of California, Berkeley.

Weerahandi, Samaradasa, Robert S. Hisiger and Victor Chien, 1994, A framework for forecasting demand for new services and their cross effects on existing services, *Information Economics and Policy* 6(2)(July):143-162.

Wolak, Frank A., 1994, Can universal access survive in a competitive telecommunications environment? Evidence from the consumer expenditure survey. a paper presented at Institut D'Economie Industrielle's (IDEI) Network Conference, Toulouse, France, October. (Mimeograph).

CHAPTER 3

Comment: Gary Madden

Telecommunications technology, and the services embodied therein, provide the mode through which information is made available to users. Communications occur so that individuals and firms can gain, satisfy needs, through access to information. This derived demand for telecommunications, is a flow, while information can be viewed either as a stock or as a flow.

Marshallian (or traditional) demand analysis considers the decision-maker's demand for flows of well specified telecommunications services. Information needs *per se* are not modelled. That is, the process by which consumers' utility functions are formed, and the manner in which communications services enter the firms' cost functions are not explicltly-modelled.

Residential and business demands are treated as separate because the interests of the decision units are fundamentally different. Residential demand for telecommunications is the outcome of calling to relevant populations of interest so as to enhance an individual's (household's) sense of well being, subject to wealth restrictions. The fundamental tool used to examine the behaviour of the consumer is the individual's static utility function, whereas, business demand for communications is undertaken for cost minimization (profit maximization) reasons by the firm. This implies a heterogeneous cost function approach; the manner in which telecommunications services enter the firm's cost function depends on the role of telecommunications inputs to the firm. For example, in the service sector telecommunications would be a direct cost, while these services are an indirect cost to manufacturing firms.

Early demand analysis, both residential and business, explain separately regional call traffic revenue and telephone network penetration rates in terms of regional income and call-price indices. This literature is well summarized in Taylor (1994). Several recent attempts to model the demand for residential service choice and the demand for calling have been completed using cross-sectional micro-level survey data.[1,2] Recent analysis of manufacturing demand for telecommunication services in the manufacturing industry is undertaken by Antonelli (1990).

When the available data (unit of analysis) is the individual decision-maker then the modelling of demand is more amenable to the Lancastrian approach to demand analysis whereby products or services are more or less well described as a bundle of measurable characteristics. This approach to demand analysis is of potentially great importance, especially in an environment characterized by rapid technological change, with associated product developments.[3]

Another and in some ways related thrust in current research, within the traditional approach, is the reconsideration of the use of a static utility function. Antonelli and others argue that individuals' utility functions are dynamic rather than static. It has been recognized, since the pioneering work of Rohlfs (1974), that utility functions defined over telecommunications services are interdependent. The full extent of the resultant network externality on demand has yet to be adequately addressed. The contribution of Antonelli is the recognition of advertising and learning as additional arguments that potentially alter preferences for telecommunications services.

The above discussion can be classified under the heading of Traditional\ Lancastrian demand analysis for extant and emerging telecommunications products that are derived from the demand for the underlying information good. The current state of this literature is subject to several types of limitations.

First, the cultural context in which telecommunications demands are formed is ignored. That is, Menou implicitly argues that preferences and the nature of communications demand critically depends on the institutional framework. The application of some standard model or set of procedures to analyse demand, in all institutional contexts, is spurious. Clearly an important research agenda item is to gain a better understanding as to the way in which cultural features of a population affect end-user demand calculations, and then with this understanding, consider whether and the means by which traditional demand analysis can be modified to incorporate cultural institutions into the analysis. This analysis may also need to include a closer examination of information needs, and how the cultural distinctions affect the relationship between communication flows and information stocks.

Second, Moyal broadens the above discussion to include gender issues. The perspective here is that women's perception and use of communication mediums are fundamentally different from those of male information users. To date little economic analysis has been focused in this direction. Clearly further research in this area needs to identify, in a systematic manner, the way in which gender leads to different end-user demands.

Third, Swann mentions that the two-way interactivity usually associated with telecommunications is irrelevant for data transportation over communications networks. He argues, within the business sector, that this one-way mode of transmission is becoming increasingly important. Menou is concerned as to whether the associated scale economies of information transfer are lost when the burden is shifted to end-users. Further, modes of information transmissions such as voice-mail, have the potential to alienate consumers. The social costs of the introduction of new technology is accordingly a proposed area for further research.

Finally, the relative efficiency of information transfer by alternative modes of transmission is raised as an issue in several places, that is the choice of optimum mix of communication modes, especially in the context of modelling the demand for new products. Issues that arise from the discussion include Weerahandi *et al.*'s (1994) concern that the market for new products should not be considered in isolation. The discussion above concerns the introduction of new communications products in isolation rather than consider this introduction among a group of competing products. The size of the "total" market and the substitutability of alternative services to provide "standard" service needs to be considered. More fundamentally de Fontenay is concerned with the ability of researchers to define new products in terms of technical and other characteristics. Also as new technologies provide a much wider range of services to end-users, the ability to clearly identify "other" or related commodities should be addressed.[4] The issue is, how to characterize new, or for that matter, old technology? Swann in chapter 4 of this volume discusses the importance of this emerging field of research.

Notes

1. See for example Train, McFadden and Ben-Akiva (1987), Train, Ben-Akiva and Atherton (1989), Taylor and Kridel (1990) and Madden, Bloch and Hensher (1993).
2. Several forum members are pessimistic about further developments in this area of research due to a lack of data.
3. See Weeranhandi, Hisiger and Chien (1994) for a seminal study in this field.
4. Madden and Simpson (1995) are undertaking research in this area for the demand for generic broadband service categories.

References

Antonelli, C., 1990, Information technology and the derived demand for telecommunication services in the manufacturing industry, *Information Economics and Policy* 4:45-55.

Madden, G., H. Bloch and D. Hensher, 1993, A model of Australian telephone network subscription and calling demands: evidence from a stated-preference experiment, *Information Economics and Policy* 5:207-230.

Rohlfs, J., 1974, A theory of interdependent demand for a communications service, *Bell Journal of Economics and Management Science* 5:16-37.

Taylor, L., 1994, *Telecommunications Demand in Theory and Practice* (Kluwer Academic Publishers, Dordrecht).

Train, K., D. McFadden and M. Ben-Akiva, 1987, The demand for local telephone service: a fully discrete model of residential calling patterns and service charges, *The Rand Journal of Economics* 18:109-123.

Train, K., M. Ben-Akiva and T. Atherton, 1989, Consumption patterns and self-selecting tariffs, *Review of Economics and Statistics* 71:62-73.

Weerahandi, S., R. Hisiger and V. Chien, 1934, A framework for forecasting demand for new services and their cross effects on existing services, *Information Economics and Policy* 6:143-162.

CHAPTER 3

Comment: David Cracknell

I believe it is possible to combine the two approaches to demand forecasting — e.g. by specifying econometric models for particular customer segments or particular call types as we have done in the UK. The key is perhaps to understand customer needs on a broad basis, and this can be used to forecast demand for new telecommunication services. As an example, Video on Demand is a substitute for Video Rentals and potential demand can be established by comparing price and other characteristics (e.g. convenience, range of choice, quality of transmission).

The answer to Menou's point about knowing people as individuals is, I believe, the development of an appropriate level of segmentation, A forthcoming paper of mine, in the *International Journal of Forecasting*, observes that the three key issues concerning the level of segmentation are that:

- It should be sufficiently fine that the customers within it could be regarded as a homogeneous group for the purpose of choosing between pricing options offered by BT and competitors;
- It should be sufficiently aggregated so as not to stretch sample based information sources beyond tolerable limits of accuracy; and
- It should not be so complex that model run times were unacceptable for pricing development.

In the context of the development of pricing options in the UK, the level of segmentation had to reflect the range of call usage, but be most finely concentrated on the users where the greatest range of choice was available. The segmentation selected is based on a number of criteria which include:

- Customer type: residential, plus five categories of business based on lines per site.
- Usage: an eleven-way segmentation of call bill per line.

This leads to a total of 66 segments (55 business, 11 residential).

CHAPTER 4

Localized knowledge percolation processes and information networks[1]

Cristiano Antonelli

1. Introduction

The notion of network is gaining momentum in social sciences: it provides in fact an appealing metaphor to analyse social events where the holistic approach and the individualistic one merge. Networks are constituted of interdependent nodes whose behaviour is influenced by the system of relations with the other nodes. More specifically networks make it possible to appreciate the effects of direct interdependence that are not fully cleared by market prices.[2] Moreover within networks interaction takes place among large numbers of players, so that each agent is not able to interact strategically with the others so as to anticipate the unexpected consequences of the network on own actions.[3]

Within networks the analysis of the action of each agent provides only a part of the story: the properties of the network into which the individual action is embedded are in fact essential to understand both the ultimate result of his or her action as well as the outcome of the aggregate action. The properties of the system are relevant in fact to understand also the result of the aggregate action of agents. On the other hand it seems appropriate to stress that the analysis of the behaviour of each agent is essential to grasp the dynamics of the aggregate: network analysis is not fully holistic as far as the interdependence between agents and the system into which their action is embedded is stressed. Finally the properties of the system can be thought likely to change over time either spontaneously because of the intentional action of agents or because of exogenous interventions such as network policies.

The spread of interest in this metaphor has been rapid in the areas of economics where the evidence about interdependence, in terms of externalities, complementarity, interrelatedness, imperfect divisibility among production and utility functions is strong. Hence regional economic, industrial organization in the theory of firm and markets, demand theory, are using more and more often the notion of network. The information and communication technology revolution supplies further stimuli to the widespread use of this metaphor for the relevance of infrastructural and virtual networks in its structure both on demand and supply side: communication networks provide one of the most clear examples of systems characterized by agents whose behaviour is strongly affected by technical and pecuniary externalities on the supply side and network externalities, both intraindustrial and interindustrial, on the demand side.

In this paper we shall try to contribute to the methodology of network economics with an exploration into the properties of networks as percolation structures into which information flows. With such a methodology we shall try to apply network economics to the economics of science and information relying on the notion of localized technological knowledge. One of the most fascinating questions an economist can try and provide an answer is why some countries are more apt to generate higher levels of technological knowledge than others, why in some countries the rate of introduction and diffusion of innovations is larger than in others and why in some countries productivity growth is larger than in others and last but not least whether new information and communication technologies can provide a new spur of opportunities to enhance the rates of introduction of technical change and productivity growth (Abramovitz, 1989).

In order to face such a problem we shall elaborate on the notion of localized technological knowledge and stress the role of communication in its generation. On such basis we shall show how the properties of innovation networks in terms of receptivity and connectivity matter in explaining the capability to generate new technological knowledge when it is based on bottom-up inductive processes of collective learning (Antonelli, 1995).

2. Localized technological knowledge

The notion of localized technological notion contrasts the Arrovian economics of knowledge based upon a deductive — top-down — theory of the relationship between science and technology where appropriability and the public character of knowledge are basic elements (Arrow, 1962, 1969). The Arrovian economics of knowledge contends that technological knowledge has the characteristic attributes of a public good:

- its use is non-excludable and non-rival;
- its production and use are characterized by high levels of indivisibility;
- once generated it can be transferred and learnt at low costs;
- it cannot be fully appropriated by innovators;
- it can be applied to an unlimited range of production processes and goods;
- it cannot be traded without disclosure hence its value is difficult to assess.

The appropriability framework rests upon four founding blocks:

(1) Firm allocation of resources to fund research and development activities is a profit-maximizing decision aimed at extracting the transient quasi-rents that are associated with the successful introduction of technological innovations.

(2) Knowledge can be produced by means of research and development expenditures in isolation by each firm which however benefits from the spillover of scientific knowledge generated from universities and other scientific institutions.

(3) Technological innovations produced by means of the blending of scientific knowledge and research and development expenditures conducted by firms can be transferred and imitated by third parties at negligible costs, that are in any case much lower than innovation costs.

(4) Returns to research and development expenditures may consequently be very low for firms although they may be very high from a social point of view. The general supply of research and development activities in a free market is thus expected to be lower than socially desirable. Hence the classic conditions of a market-failure arise: markets in fact are not able to allocate the Pareto-equilibrium quantities of resources in the generation of technological innovations.

A growing number of authors recently challenged the traditional view about technology as a pure public good and argue that technology has a strong systemic character so that each unit of technological knowledge can be created and traded only within a larger framework including the array of complementary and interrelated units of knowledge generated by other firms. Therefore each unit of technological knowledge is largely excludable and its use partly rivalrous. In fact technology tends to be firm-specific and it is costly if not impossible to use elsewhere. Communication plays a major role in this context:

(i) the transfer and adaptation of technological knowledge from one firm to another involves specific costs that according to many empirical analyses amount to an order of magnitude very close to the cost of first introduction (Nelson, 1990; Rosenberg, 1990);

(ii) in turn however the generation of localized knowledge is viewed as the outcome of a collective undertaking strongly influenced by the availability of information and communication channels among learning agents (Von Hippel, 1988, Allen, 1983).

The divide between the standard views and the neo-technological ones applies also to the production of technological knowledge. In the standard approach the generation of technological knowledge is the outcome of institutionalized efforts geared towards the introduction of innovation that take place in a deductive chain that draws from scientific discoveries and general methodological procedures elaborated mainly in academia, easily understood by all potential users.

In the neo-technological one the generation of new knowledge is mainly the outcome of the interaction of the non-institutionalized efforts of innovators that draw on learning processes highly localized and specific to the history and experience of each innovator. Localized technological knowledge consists of highly specific and tacit knowledge with strong specific and idiosyncratic characters. Localized technological knowledge emerges from daily routines and from the tacit experience acquired in using capital goods, in producing and manufacturing, in interacting with customers, with other manufacturers. Research and development expenses defined as the resources allocated in the specific activity of experimenting and developing new products and new processes are only an aspect of a more general process of learning and capitalizing on the experience acquired. Localized technological knowledge can also be viewed as the product of a systemic bottom-up process of induction from actual experience which contrasts sharply the individualistic top-down process of deduction from general scientific principles on which the received theory of knowledge as a public good rested.

A new wave of empirical and theoretical works have recently paid increasing attention to the systemic character of the generation of technological knowledge (Freeman-Soete, 1990; Romer, 1986). The new evidence shows that innovating firms cannot be analysed in isolation: innovation capability depends in fact also on the amount of information that each firm is able to receive from the environment into which it operates. The traditional chain of factors that links innovation capability exclusively to intramural R&D expenditures is consequently rejected and together with learning capabilities increasing attention has been recently paid in this context to technological externalities spilling from the learning and research activities of other firms. Technological externalities are defined as the direct — not mediated by market transactions — effects on the technological capacity of each firm produced by complementarity and interrelation of R&D activities and technological learning among firms. Hence technological externalities spill from complementarity and interrelatedness among firms in the generation of new technologies rather than complementarity and interrelatedness of the actual production processes in use (Griliches, 1979, 1984, 1988; Jaffe, 1986, 1989; Katz-Shapiro, 1986; Arora-Gambardella, 1990; Carlsson-Stankiewitz, 1991).

A large body of empirical research shows that: i) the process of introduction of innovations of each firm interacts in such a way as to condition each other, ii) the introduction of a single technological innovation is highly conditioned by decisions regarding the adoption of complementary or related innovations by other firms; iii) the close interaction with customers and the assistance to them in learning how to use the new products generates important additional knowledge about the capability of the new products that reflective sellers are able to appropriate and consequently elaborate upon. (Allen, 1983; Von Hippel, 1988; Lundvall, 1985); iv) explicit cooperation among firms plays a major role in assessing the innovative capability for it enhances the circulation of information; the opportunities for external learning — that is learning from the specific knowledge generated by other firms; the scope for capitalizing on potential complementarities between the variety of firms and between the different R&D activities performed by each firm (Teece, 1986; Cohen-Levinthal, 1989; Klein, 1992; Malerba, 1992; Nelson, 1993).

New information and communication technologies provide large evidence on the systemic character of technological change: information and communication technologies are themselves the outcome of the convergence of a variety of complementary innovations generated in a wide range of industries and firms ranging from telecommunications, informatics, micro-processors, optoelectronics, space industries. The generation of each of these innovations is strongly influenced by the availability of complementary technological information as well as their successful application is very much influenced by the availability of complementary innovations (Antonelli, 1992; Mansell, 1994; Rosenberg, 1994).

In sum, the capability to innovate successfully appears to be strongly conditioned by learning opportunities and by the accumulation of specific knowledge that are both internal and external to each firm. In so doing it seems possible to highlight the strong systemic character of technological change and the role of externalities and interdependence in the introduction of a new technology.

Technological knowledge now is viewed as the product of a bottom-up systemic process of induction from actual experience and communication which contrasts sharply with the top-down process of deduction from general scientific principles on which the received theory of knowledge as a public good rested. The approach to localized technological knowledge as a quasi-private good elaborates on recent understanding about the origins and flows of the generation process of technological knowledge within organizations and in the relations among organizations:

(1) technological knowledge is embedded in the "circumstances" in which the firm operates, hence technological change is localized in the techniques currently used by each firm, in the markets in which each firm operates, in the existing information channels among firms and customers, in the organizational structure of firms, in the informational space in which each firm operates.

(2) The generation of technological knowledge is the result of a joint process of production, learning and communicating of which research and development activities should be only considered as a part. Consequently research and development activities cannot and should not be considered the sole factors in the generation of new technological knowledge and should not be separated from the current flow of activities within the firm and in the relations between the firm and its environment.

(3) Flows of goods and flows of knowledge cannot be separated. In fact the capability of firms to generate technological innovations rests upon the continued effort to keep open all the information and communication channels that parallel the flow of goods and to capitalize upon the opportunity to learn they provide by creating an appropriate organizational environment (David, 1993a, 1993b). Localized technological knowledge emerges also from the daily interaction of learning firms: in such interaction communication and trading of information play a major role.

Hence the study of the determinants of the innovation capability of an economic system must take into account two classes of factors: i) the amount of resources devoted to generate innovation by each agent in the system and its receptivity to technological knowledge generated by third parties; ii) the properties of the system in terms of connectivity and distribution of receptive agents.

3. Innovation and percolation processes

Let us formalize the character of the innovation generation process we have discussed above. According to the hypotheses outlined we assume that for each firm the production function of technological knowledge can be formalized as the result of the interaction of internal research and learning activities and the access and absorption of technological knowledge spilling from other firms that are part of the same innovation system:

$$T_i = f(R\&D\&L_i, g(C, R\&D\&L_{n-1})) \quad (1)$$

where T_i is the technological knowledge produced by the firm i; R&D&L is the amount of resources devoted to learning and research and development activities; C is the extent of communication and absorption of the innovation activity of each firm with the other n firms that belong to the economic system into which the firm i operates; and g measures the effects on the productivity of intramural resources of the external learning.

Hence the amount of technological knowledge an economic system is able to generate will be determined by the amount of resources devoted to fund research, development and learning activities and the extent to which effective communication takes place. Effective communication requires that two quite distinct events coincide:

(i) the emission of information; and

(ii) the reception of information.

Percolation processes have been studied in physics as the outcome of two classes of forces termed as connectivity and receptivity. The percolation probability is measured by the combined result of receptivity probability and connectivity probability. It seems possible to use the methodology of percolation processes to study the probability that within innovation networks the flows of effective communication and exchange of information take place and actually implement the productivity of the research and learning efforts of each firm in the system. In fact the empirical evidence shows that the distribution of transmission links among firms, that is the connectivity coefficient of a network, varies and the extent to which each firm is able to absorb the amount of information transmitted, that is the receptivity coefficients, is distributed unevenly in the system. Connectivity coefficients can be lower than n-1, the maximum number for a system with n firms, and receptivity levels can be lower than 100%, when all firms are able to absorb all the information transmitted. Hence we can spell out the hypothesis that the innovation output of an economic system depends upon the amount of resources invested in research and learning activities and the properties of percolation process which is in place within innovation networks.

With the notion of percolation probability we refer here to the probability that each firm i can effectively communicate elements and bits of technological information from the other firms that are part of the same system and is able to receive them. The percolation probability qualifies the probability that information flows, like movements of particles through a structured medium, constituted of homogeneous or non-homogeneous elements, reach all the elements of the system and are absorbed (David-Foray, 1994).[4] The structured medium can be thought of as a network constituted of transmission links and nodes: some of the nodes can be more receptive than others and some transmission links can be more effective than others.

The percolation probability is determined by the properties of the system in terms of quality of the connectivity and quality of the receptivity where the former refers to the quality of the information and communication channels among nodes and the latter to the extent to which there are technological externalities, that is effectively complementarity among firms with respect to the generation of technological knowledge. The flow of information among units is stopped by low quality connections or low quality receptivity. The larger the flow of complementary information the larger the generation levels of technology by each firm and consequently by the system.

The basic properties of the system therefore appear to consist in the actual complementarity in generating new technological knowledge that is traduced into receptivity and the extent to which each firm is able to interact and participate in the information flow with the other firms that is traduced in terms of connectivity.

Let us define "G" as a network which consists of a set of nodes "O"

(2) $$O = (o_1, o_2 \ldots o_n)$$

each of which is connected to the others by a set of transmission links "T"

(3) $$T = (t_1, t_2 \ldots t_n).$$

An operative path in G from a firm F_i to a firm F_j in a finite sequence of the form $(t_{12}\ F_2\ t_{23}\ o_3 \ldots t_{n-1}\ F_n)$ where t_{ij} denotes a relational line connecting F_i to F_j. The network G is connected if, for each pair of firms F_i and F_j, there is a path in G from F_i to F_j.

We now construct a random maze on G as follows. Let each organization o of G be "open", or responsive to the influence of any of its neighbour's innovative action, with probability p_o, or closed (unresponsive to local technological externalities) with probability $q_o = 1\text{-}p_o$. We shall call such probability the receptivity probability. Similarly each interorganizational transmission link t attains some minimal or threshold density of interactions between i and j (which is sufficient to influence decisively the ith and jth firms' level of innovation output) with probability At or it fails to do so with probability $q_t = 1\text{-}p_t$. We can call such probability the connectivity probability. Moreover we shall assume all these events are to occur independently of each other.

An operative path d between F_i and F_j is defined as follows:

(4) $$d = (t_{12}\ F_2\ t_{23}\ o_3 \ldots t_{n-1}\ F_n).$$

The operative path between F_i and F_j is said to be "open" if all its communication links attain the minimum sufficient density and all the firms are influenced by technological externality in their innovative activity. Thus the probability that the particular path d is "operational" in that sense is given by $(p_o\ p_t)^{n-1}$.

The percolation probability $P(p_o, p_t, G)$ is the probability that the research and learning activities carried out by each firm can implement the technology T generated by some infinite set of nodes in G. If we suppose now that G is fixed and G is infinite we can simplify the percolation probability as follows:[5]

(5) $$P(p_o, p_t, G) = P\ (p_o, p_t) = P.$$

Clearly the mixed percolation probability P is a non-decreasing function of p_o and p_t and $P(0, 0) = P(1, 0) = P(0, 1) = 0$. Thus, $P_o(p) = P(p_o\ 1)$ and $P_t(p) = P(1\ p_t)$ are respectively the node percolation and the connection probabilities of this network.

A fundamental mathematical property of the percolation process is that there are some critical values $p_o > p_o^*$ and $p_t > p_t^*$ beyond which there will be a positive probability that percolation occurs, and below which the percolation probability is zero.

In other words the system undergoes a system "phase transition" when these underlying critical probabilities are attained. There are corresponding critical values at which the node percolation and the transmission percolation probabilities respectively become positive. These define the endpoints of a region above which a "mixed percolation process" (one for which it is not certain that either all nodes or all links of the network are open) will have positive probability of achieving complete percolation (David-Foray, 1994).

The receptivity of the system is more "brittle" than the connectivity. The following general property is important:

(6) $$P_o(p) < P_t\ (p) \quad \text{whenever } 0 < p < 1.$$

This property suggests that technological information percolates more easily via an imperfect set of connections between receptive nodes than through a completely connected network whose nodes are imperfectly receptive.

Another generalization states that:

(7) $$P(p_o\ b, p_t) < P(p_o, p_t\ b) \quad \text{whenever } 0 < p_o < 1, \quad 0 < p_t < 1, \quad 0 < b < 1.$$

This inequality suggests that it is more effective to modify, by a factor b the receptivity of members of a population than to modify the connectivity with efforts of the same amount.

We can fully articulate our hypothesis and define the maximum attainable efficiency of innovation systems as follows:

(8) $$T_S = (S\ R\&D\&L)^{((n-1)\ gPP)},$$

where T_S is the amount of technological knowledge produced in the system, g is the effect of effective information arrived and received to each firm in the system on its own innovative efficiency and PP the percolation probability.

It is clear that the mixed probability of the percolation process has a strong effect on the total amount of resources devoted to research and learning activities. More specifically we see that, according to the properties of percolation systems:

(i) the larger the receptivity of the research agenda of each firm to the information transmitted, the larger the innovative output;

(ii) the larger the connectivity among firms, the larger the innovative output, although to lesser levels with respect to the effects of receptivity coefficients;

(iii) for levels of connectivity and receptivity coefficients that are $p_o < p_o^*$ and $p_t < p_t^*$ lower than respectively the minimum connectivity and receptivity thresholds the general efficiency of the innovation system is severely damaged. Within the network some areas can experience connectivity and receptivity levels that are larger than the thresholds so creating "maculated" innovation networks.

The model presented so far assumes global interaction in that all the agents have the same probability of networking, that is that all the agents have the same number of connections. But interaction can be localized in that agents have a finite number of connections that is smaller than n-1. A hierarchy of agents, whose productivity depends on the number of connections is likely to emerge in a system characterized by localized interaction. In such a context it can be assumed that each connection requires an effort to be established: hence connections become the outcome of intentional action and can be considered as endogenous. Moreover the levels of funds invested in R&D activities can also be thought to be influenced by the number of connections that in turn influence their productivity levels in terms of innovative output. Conversely, however, firms with given amounts of financial resources available to get some desired levels of innovative output might be induced to substitute internal R&D activities with connections.

In a dynamic process we see that the larger the number of connections and the more effective the connections in terms of communications, the larger the innovative output, hence the profits, consequently the larger the amount of resources devoted to R&D activities and to build connections. These results have important implications for technology policy: all efforts to increase the percolation probability of innovation systems are in fact likely to have strong positive effects on its innovative efficiency for given levels of resources dedicated to learning and research activities. A new agenda for technology policy can be built on these bases.

4. Conclusions

The traditional argument that technological knowledge, grafted into blueprints is on the shelf, as elaborated on the basis of the Arrovian economics of knowledge, represents the pillar on which a large portion of the economics of knowledge is built (Arrow, 1962, 1969). The assignment of property rights, such as patents, reduces only to a limited extent the difficulty for innovating firms to appropriate all the returns to technological knowledge. Consequently technology is likely to be undersupplied. Traditionally technology policy has been consequently based on incentives to push firms to fund R&D activities beyond the levels suggested by private marginal productivity in order to take into account the social positive returns. The recipe of technology policy based on the traditional view consists of: i) Special institutions such as patents, that is a special class of property rights, should be created, enforced and assigned to innovators in order to secure the necessary levels of returns to their research and development expenditures; ii) Large firms with high levels of market power, prior to innovative efforts, should also be favoured in order to make longer the duration of monopolistic powers that in turn make possible the "indirect" appropriation of the stream of quasi-rents associated with the introduction of technological innovations; iii) Direct State interventions, like research and development subsidies, are required in order to reduce the costs for firms of the activities that are necessary to generate technological innovations (Stoneman, 1983, 1987).

These traditional views are now contrasted by a new approach. In this new approach technological knowledge has a strong localized character. Localized technological knowledge is very much the outcome of a systemic and inductive, bottom-up, process of generation of new knowledge that emerges from: i) the experience acquired by agents in daily routines; ii) the access to and the absorptive capability of external technological information generated by the other members of the system; and iii) the competence they have been able to build upon. Localized technological knowledge is the outcome of a path-dependent dynamic process in that it reflects the original — accidental — characters of the economic system in which the dynamic process starts as well as the intentional behaviour of learning agents whose innovative conduct is fully embedded in the economic system they are part of (David, 1975, 1993a).

The divide between the standard approach and the contrasting one concerns both the supply and the demand of technological knowledge. The standard approach stresses the pure informational aspect of technological knowledge and views it as the outcome of a scientific process, i.e. of the application of general rules to a specific problem that can be consequently and almost costlessy imitated and applied by everybody else. The contrasting approach stressing the technological and communication aspect of technological knowledge in fact focuses attention on the tacit character of the know-how necessary both to generate and to use it and on the role of communication and exchanges of information among parties involved in the generation of tacit knowledge. Specificity of technological knowledge to the environment that generates it and communication requirements are the two sides of the same coin. So far the notion of localized knowledge seems to retain and develop in many respects the notion, put forward by Hayek (1976), of "catalysis", that is the inductive emerging of knowledge and information from the daily interaction and experience of economic agents.

The attributes of the process that leads to the generation of localized technological knowledge make it possible to use the methodology of percolation processes to understand its dynamics. The analogy between percolation processes and innovation ones rests on: i) the structured system into which the process takes place; ii) the complementarity of technological information that is relevant to generate technological knowledge; iii) the role of communication channels among firms.

The relevance of the percolation process in answering the basic questions about the determinants of the uneven distribution of innovation capability across countries and regions and industrial systems we raised is manifold:

(i) it shows convincingly that because of the localized character of technological knowledge, the generation of new knowledge is strongly influenced by the quality of communication flows among firms as well as by the complementarity of the information acquired by each firm in its tacit learning process as measured by percolation probability;

(ii) it provides important insights into the properties of innovative processes: the outcome of percolation processes can be predicted if the levels of the two basic characters of the percolation process, such as connectivity and receptivity are known;

(iii) hence a technology policy can be built by elaborating on the parameters that govern the levels of connectivity and receptivity of the system.

The cumulative engine that percolation processes describe are likely to be very sensitive to the values of p_o* and p_t*. Systems where respectively complementarity or receptivity and connectivity fall below these threshold levels are likely to be unable to grab the positive cumulative effects activated by percolation processes on the generation of localized technological knowledge. Hence architectural issues become relevant not only to understanding the long terms dynamics of innovation and growth: system design that is able to implement the appropriate properties can play a major role in technology policy (Sah-Stiglitz, 1986 and Mokyr, 1990).

For the same reason we can articulate the hypothesis that percolation processes provide an important rationale for understanding the strong stability over time of the technological advance and the specialization of countries. For given levels of resources invested in research and learning activities the amount of technology generated is likely to be larger in some sub-elements of the industrial systems where a strong complementarity and strong information flows are already in place (Archibugi-Pianta, 1992). Hence not only the levels of technological capability, in terms of rates of introduction of technological innovations, but also the technological specialization in terms of industrial and technological direction across sectors and products, is likely to be strongly path-dependent. Hence we can expect that:

(i) New industrializing countries where communications channels and complementarity are yet to be established and organized are likely to experience much higher innovation unit costs than industrialized countries where a long lasting practice of cooperation and actual complementarity has been fully embedded in industrial systems. When we take into account the possibility of interactions between productivity levels and the properties of innovation networks the dynamics of percolation seem to provide an important explanatory framework to understand cumulative self-sustained growth led by learning to interact within the network. Innovating countries with well designed innovation systems are in fact likely to experience over time a continued increase in their innovation productivity levels provided that positive feed-backs between the quality of connectivity and receptivity and the levels of innovation productivity take place either "spontaneously" or because of technology policies. Successful agents within innovation networks can learn to communicate both in terms of connectivity and receptivity as soon as they realize that their innovation productivity is positively influenced by the communication network (Stiglitz, 1987).

(ii) Countries with high levels of technological capability in a given technological area such as the engineering industries in Italy are likely to experience a continued advance in these areas with slight changes over time in related areas such as the new emerging mecatronics generated by the merging of electronics and engineering. Hence we can expect strong path dependence effects in the evolution of the technological specialization of countries and firms. With given levels of external pressure such as changes in relative prices of production factors countries with a different percolation structure and hence a different structure of their innovation systems will react differently. More specifically it seems that the hypothesis of the international life cycle of technology according to which countries should follow a well defined evolution in their industrial composition of labour intensive productions to capital intensive ones should be re-assessed in order to take into account the hysteric character of the evolution of technological specialization as determined at each point in time by the actual levels of connectivity and receptivity. Therefore countries are more likely to implement their own technological specialization and to experience continued increases in the productivity levels around a given initial composition of their industrial structure. Countries specializing in consumer products are more likely to experience continued growth in their efficiency in producing consumer goods rather than eventually specializing in engineering industries.

The dynamics of percolation systems, as applied to understanding the dynamics of innovation systems, has important implications for technology policies. Here the useful distinction introduced by Ergas (1987) between innovation policies and diffusion policies can be stretched so as to articulate the distinction between network-technology policies aimed at increasing the efficiency of innovation efforts by means of interventions on the degree of coherence of innovation networks and technology policies aimed at increasing the amount of innovative efforts of firms. Hence we expect that:

(i) Countries where more systemic network-approaches to industrial policy can experience higher rates of technological advance than countries with a more individualistic approach, for given levels of resources invested in research and learning activities;

(ii) All efforts devoted to strengthen the complementarity of the research and learning activities of the firms that belong to the system have high chances to generate important effects in terms of larger amounts of technology generated by each firm and by the system. *Ex ante* cooperation schemes in long-term technology projects are likely to provide useful guidance to firms in order to plan their efforts in a complementary way.

(iii) Cooperation among innovating firms is likely to generate even stronger results in terms of effects on connectivity. The creation and implementation of effective information networks among firms that enhance the levels of connectivity is likely to have strong positive effects as well as all the interventions that favour actual complementarity of research and learning activities among firms.

(iv) More generally traditional technology policies based upon incentives schemes finalized to increase the levels of R&D could be reoriented in order to take into account the properties of the innovation systems such as connectivity and receptivity.

(v) The introduction of meta-standards that increase the complementarity and compatibility of research agenda of each firm can contribute to increasing the connectivity of innovation networks. *Ex ante* standardization can become an important tool of technology policy (David-Greenstein, 1990).

(vi) A re-assessment of property rights seems necessary: if property rights are successfully applied and appropriation is higher in fact a clear problem of reduction of communication flows among firms emerges and with it an issue of underutilization of the innovative capability of the network arises with a clear trade-off between static efficiency and a dynamic one. Systems that provide lower levels of property rights have in fact higher chances to stimulate the flow of information exchanges among firms and consequently to obtain higher rates of return from the resources invested in R&D activities (Ordover, 1991).

(vii) New information and communication technologies seem to provide important opportunities to increase the connectivity of innovation networks and consequently the productivity of resources used in research and learning activities, provided that open network architectures are implemented so as to maximize inclusion. Hence telecommunication policies and innovation policies appear areas of increasing overlapping and complementarity.

Notes

1. The comments of Peter Swann and Jean Michel Dalle to preliminary versions of this paper are acknowledged as well as the support of the national and local MURST research funds.
2. Within markets in fact interdependence takes place all the time, but it is fully mediated by prices.
3. When numbers are sufficiently small strategic direct interaction can be taken into account by means of game theoretic approaches.
4. In the formal exposition of the mathematical properties of percolation processes we follow closely David-Foray (1994).
5. Actually the percolation probability depends also on the levels of external pressure, denoted as R so that the full percolation probability equation reads as follows:

   (5.1) $$P(p_o, p_t R, G) = (p_o, p_t) = P.$$

   Equation (5.1) can be reduced to equation (5) with given levels of R and G. In our case in fact we have not exogenous factors (David-Foray, 1994).

References

Abramovitz, M., 1989, *Thinking about Growth* (Cambridge University Press, Cambridge).

Allen, R.C., 1983, Collective invention, *Journal of Economic Behavior and Organization* 4:1-24.

Antonelli, C., ed., 1992, *The Economics of Information Networks* (North-Holland, Amsterdam).

Antonelli, C., 1995, *The Economics of Localized Technological Change and Industrial Dynamics* (Kluwer Academic Press, Boston).

Archibugi, D. and M. Pianta, 1992, *The Technological Specialization of Advanced Countries* (Kluwer Academic, Boston).

Arora, A. and A. Gambardella, 1990, Internal knowledge and external linkages: theoretical issues and an application to biotechnology, *Journal of Industrial Economics* XXXVIII:361-380.

Arrow, K.J., 1962, Economic welfare and the allocation of resources for invention, in: R.R. Nelson, ed., *The Rate and Direction of Inventive Activity: Economic and Social Factors* (Princeton University Press, Princeton).

Arrow, K.J., 1969, Classificatory notes on the production and distribution of technological knowledge, *American Economic Review* 59:29-35.

Buchanan, J.M., 1965, An economic theory of clubs, *Economica* 34:1-14.

Carlsson, B. and R. Stankiewitz, 1991, On the nature function and composition of technological systems, *Journal of Evolutionary Economics* 1:93-118.

Cohen, W.M. and D.A. Levinthal, 1989, Innovation and learning: the two faces of R&D, *Economic Journal* 99:569-596.

D'Aspremont, C. and A. Jacquemin, 1988, Cooperative and non-cooperative R&D in duopoly with spillovers, *American Economic Review* 78:1133-1137.

David, P.A., 1975, *Technical Choice Innovation and Economic Growth* (Cambridge University Press, Cambridge).

David, P.A., 1987, Some new standards for the economics of standardization in the information age, in: P. Dasgupta and P. Stoneman, eds, *Economic Policy and Technological Performance* (Cambridge University Press, Cambridge).

David, P.A., 1993a, Knowledge property and the system dynamics of technological change. *Proceedings of the World Bank Annual Conference on Development Economics* (The World Bank, Washington).

David, P.A., 1993b, Path-dependence and predictability in dynamic systems with local network externalities: a paradigm for historic economics, in: D. Foray and C. Freeman, eds, *Technology and the Wealth of Nations* (Pinter Publishers, London).

David, P.A. and S. Greenstein, 1990, The economics of compatibility standards: an introduction to recent research, *Economics of Innovation and New Technology* 1:3-42.

David, P.A. and J. Rosenbloom, 1990, Marshallian factor market. externalities and the dynamics of industrial localization, *Journal of Urban Economics* 28:349-370.

David, P.A. and D. Foray, 1994, The economics of EDI standards diffusion, in: G. Pogorel, ed., *Global Telecommunications Strategies and Technological Changes* (North-Holland, Amsterdam).

Ergas, H., 1987, The importance of technology policy, in: P. Dasgupta and P. Stoneman, eds, *Economic Policy and Technological Performance* (Cambridge University Press, Cambridge).

Freeman C. and L. Soete, eds, 1990, *New Explorations in the Economics of Technological Change* (Pinter, London).

Griliches, Z., 1979, Issues in assessing the contribution of research and development to productivity growth, *Bell Journal of Economics* 10:92-116.

Griliches, Z., ed., 1984, *R&D Patents and Productivity* (NBER, Chicago).

Griliches, Z., 1988, Productivity puzzles and R&D: another no explanation, *Journal of Economic Perspectives* 4:9-22.

Hammersley, J.M. and D.J. Welsh, 1980, Percolation theory and its ramification, *Contemporary Physics* 21(6).

Hayek, F.A., 1945, The use of knowledge in society, *American Economic Review* 35:519-530.

Hayek, F.A., 1976, *The Mirage of Social Justice* (University of Chicago Press, Chicago).

Hirschman, A.O., 1958, *The Strategy of Economic Development* (Yale University Press, New Haven).

Jaffe, A.B., 1986, Technological opportunity and spillover of R&D: evidence from firms patents, profits and market value, *American Economic Review* 76:984-1001.

Jaffe, A., 1989, Real effects of academic research, *American Economic Review* 89:957-70.

Katz, M., 1984, An analysis of cooperative R&D, *Rand Journal of Economics* 17:527-543.

Katz, M. and C. Shapiro, 1985, Network externalities competition and compatibility, *American Economic Review* 75:424-440.

Katz, M. and C. Shapiro, 1986, Technology adoption in presence of network externality, *Journal of Political Economy* 94:822-841.

Klein, B.H., 1992, The role of positive sum games in economic growth, in: F. Scherer and M. Perlman, eds, *Entrepreneurship Technological Innovation and Economic Growth* (University of Michigan Press, Ann Arbor).

Jacquemin, A., 1985, *Selection et Pouvoir dans la Nouvelle Economie Industrielle* (Cabay-Economica, Paris).

Lamberton, D., ed., 1971, *Economics of Information and Knowledge* (Penguin Harmondsworth).

Lundvall, B.A., 1985, *Product Innovation and User-Producer Interaction* (Aalborg University Press, Aalborg).

Malerba, F., 1992, Learning by firms and incremental change, *Economic Journal* 102:845-859.

Mansell, R., 1994, *The New Telecommunications* (Sage, New York).

Mansfield, E., 1985, How rapidly does new industrial technology leak out, *Journal of Industrial Economics* XXXIV:217-223.

Mansfield, E., M. Schwartz and S. Wagner, 1981, Imitation costs and patents: an empirical study, *Economic Journal* 91:907-918.

March, J.C. and H.A. Simon, 1958, *Organizations* (John Wiley & Sons, New York).

Marshall, A., 1920, *Principles of Economics* (MacMillan, London).

Metcalfe, J.S., 1993, The economic foundation of technology policy: equilibrium and evolutionary perspectives, in: Stoneman P., ed., forthcoming.

Mokyr, J., 1990, *The Levels of Riches* (Oxford University Press, Oxford).

Nelson, R.R., 1959, The simple economics of basic scientific research, *Journal of Political Economy* 67:297-306.

Nelson, R.R., 1982, The role of knowledge in R&D efficiency, *Quarterly Journal of Economics* 97:453-470.

Nelson, R.R., 1987, *Understanding Technological Change as an Evolutionary Process* (North-Holland, Amsterdam).

Nelson, R.R., 1990, Capitalism as an engine of progress, *Research Policy* 19:193-214.

Nelson, R.R., ed., 1993, *National Systems of Innovation* (Oxford University Press, Oxford).

Ordover, J.A., 1991, A patent system for both diffusion and exclusion, *Journal of Economic Perspectives* 5:43-60.

Richardson, G.B., 1972, The organization of industry, *Economic Journal* 82:883-896.

Romer, P.M., 1986, Increasing returns and long run growth, *Journal of Political Economy* 94:1002-1037.

Rosenberg, N., 1976, *Perspectives on Technology* (Cambridge University Press, London).

Rosenberg, N., 1982, *Inside the Black Box: Technology and Economics* (Cambridge University Press, London).

Rosenberg, N., 1990, Why do firms do research (with their own money)?, *Research Policy* 19:165-174.

Rosenberg, N., 1994, *Exploring the Black Box* (Cambridge University Press, Cambridge).

Sah, R. and J.E. Stiglitz, 1986, The architecture of economic systems: hierarchies and polyarchies, *American Economic Review* 76:716-727.

Saviotti, P.P. and J.S. Metcalfe, eds, 1992, *Evolutionary Theories of Economic and Technological Change* (Harwood Academic Publishers, Chur).

Scitowsky, T., 1954, Two concepts of external economies, *Journal of Political Economy* 62:143-151.

Simon H.A., 1982, *Models of Bounded Rationality* (MIT Press, Cambridge).

Stiglitz J.E., 1987, Learning to learn localized learning and technological progress, in: P. Dasgupta and P. Stoneman, eds, *Economic Policy and Technological Performance* (Cambridge University Press, Cambridge) 123-153.

Stoneman P., 1983, *The Economic Analysis of Technological Change* (Oxford University Press, Oxford).

Stoneman P., 1987, *The Economic Analysis of Technology Policy* (Clarendon Press, Oxford).

Teece, D.J., 1986, Profiting from technological innovation, *Research Policy* 15:285-305.

Von Hippel, E., 1988, *The Sources of Innovation* (Oxford University Press, London).

Winter, S.G., 1987, Knowledge and competence as strategic assets, in: D.J. Teece, ed., *The Competitive Challenge* (Ballinger, Cambridge).

Young, A., 1928, Increasing returns and economic progress, *Economic Journal* 38:527-542.

CHAPTER 4

Comment: Richard Joseph

Antonelli sets out an alternative view of technological knowledge, a view which he claims contrasts with the traditional arguments about this topic. The traditional argument is based on Arrow's perspective on the economics of knowledge. In this approach, "the assignment of property rights, such as patents, reduces only to a limited extent the difficulty for innovating firms to appropriate all the returns to technological knowledge". As result, technology is likely to be undersupplied and this gives rise to certain aspects of technology policy as we know it: incentives for R&D; patent legislation; and support for large firms since they may be able to capture better the benefits of their R&D.

Contrasting the "top-down" traditional approach where technology flows from science, Antonelli postulates that technological knowledge has a strong localized character. The implications are that important factors become experience, competence, ability to learn and the capability to absorb information. As far as the process of generating technological knowledge goes, the role of communication and exchange of information gains more significance than in the traditional approach. For Antonelli, he uses the methodology of "percolation" processes to explain the dynamics of the generation of new technological knowledge. This means that the overall system in which communication occurs is important as well as the existence of complementary information and communication channels among firms.

The conclusions which can be drawn from this contrast add some fresh perspectives to technology policy. As Antonelli notes, technology policy could now elaborate on the levels of connectivity and receptivity of the system as opposed to the traditional approach which emphasizes appropriability and support for R&D. This viewpoint leads Antonelli to make some observations, all of which are suggestive of areas for further research. In short, Antonelli argues that in countries where systemic network approaches to industry policy exist and where firms are able to learn and partly cooperate in innovation, there is the possibility of higher rates of technological advance. Antonelli also stresses that innovation has much to do with communication and information flow and that new information and communication technologies themselves may enhance learning within firms. While these observations have been cited in the literature before, Antonelli's work clearly establishes the case that existing technology policies are in need of revision and revitalization.[1]

It is from these observations that I believe that a fruitful research agenda can be identified from the following three themes: the issue of countries modelling each other in technology policy; technological innovation as a communication process; and the role of information and telecommunications technologies in mediating innovation.

Modelling of technology policies

Antonelli remarks that some countries with systemic approaches to industrial policy may be better off. One difficult problem facing countries wanting to improve their industry and technology policies is what to copy from apparently successful countries. More sophisticated approaches to innovation may describe the process better but they still need to provide a guide for policy action. It is at this point that political interests and the powerful symbolism often associated with new technology become important. In the practice of technology policy, there is a strong assumption that what appears to be working in one country can be easily transferred to another country to produce the same results. It is believed that not only new technology and systems of innovation, but also the public policy measures that support these systems, can be effectively transferred. This assumption is very attractive to governments looking for a quick-fix solution to development problems but at the deeper level, public policy may be doing little to create these vital innovation systems and perhaps may even damage the fragile information networks that exist in a country. This is a problem that seems to haunt technology policy.

During the 1970s and early 1980s it became popular for governments to focus on high technology growth. Typical key technologies were microelectronics and biotechnology and public money was invested in establishing embryonic high technology industries in many countries. As a broader model, the success of Silicon Valley in the US meant that other countries wanted to replicate the phenomenon. Technology parks became a popular technology policy measure to attract new and existing high technology firms to locate in a region. The result has been far less spectacular than the high technology pundits predicted a decade ago. Technology parks were conceptualized on the linear model of innovation (reflecting Antonelli's traditional approach which he criticizes).[2] The complex information flow aspects of high technology were ignored or poorly dealt with through technology parks. While technology policy in this way can be regarded as failing, it may still achieve a degree of short-term political success as policy-makers have to be seen to be doing something. The same criticism is relevant to the current interest in copying the National Information Infrastructure (NII) initiative in the US. The term Global Information Infrastructure is now popular and countries such as Canada, Japan and Australia to name but a few, have active policies to promote the so-called "information superhighway". It may be that some countries will be in for a disappointment similar to the high technology era as they attempt to model technology policies on the leading countries with little success. The need for adopting simple policy responses based on inadequate models of the innovation process must be given some of the blame.

Antonelli's observation that some countries with better network approaches to industrial policy will be better off is important here. As a valuable area for further research, I believe that we should be looking at not only how technology can be transferred effectively across countries but also how technology policy measures (such as technology parks) can be transferred and made more effective. Antonelli's observation suggests that we should be very circumspect about "quick-fix" solutions — the traditional viewpoint, on its own, is no longer adequate for guiding the policy and technology transfer process. Having said this, two further problems can be immediately identified. The first is the difficulty of linking Antonelli's insights about innovation and meshing that into a policy language that can be incorporated into technology policy. Given the current dominance of orthodox economic thinking, with its neglect of information, this work will probably require much further research and is likely to meet with some resistance. The second problem relates to the process of policy modelling itself. How does it happen, why does it happen and through what channels does it occur? Answers to these questions will throw some light on why some countries are better than others at generating both technology and effective technology policies.[3]

Technological innovation as a communication process

Antonelli makes the important observation that technological innovation should be thought of as a communication or information flow process. I believe that this observation provides the basis for some areas of fruitful further research. At a general level, the complexities of technological change have long been recognized by historians of technology and economic historians.[4] More recently, new developments in the sociology of technology have been providing insights into the complex process of innovation and diffusion.[5] I believe that there is scope for a much closer cross-fertilization of ideas across these disciplinary boundaries and it may be through the notion of information flow and communication in innovation that some bridges can be built. Of course, the important role of institutions cannot be neglected and there would appear to be scope for analysing features like intra- and inter-organizational communication patterns as part of the innovation process.

More specifically, viewing innovation as a communication process raises a number of research questions. First, as technology acquisition practices change within firms there will be a growing need to investigate what communication patterns are occurring, for example, between firms in strategic alliances or say between firms linked by a cooperative R&D project. Understanding these communication links should provide considerable insights into the innovation process. Second, networked communities of scientists, engineers and technologists should not be ignored either as there is scope for a better understanding of the boundaries between activities that are arbitrarily labelled by science policy-makers as "science" and "technology". Third, links between government, industry, universities and research agencies also need further investigation since it is through these links that many technology policy initiatives in vogue today are based. In short, once innovation is considered to be an information flow issue then a range of new research dimensions are immediately opened up.

The role of information and communications technologies in mediating innovation

Antonelli makes the point that new information and communication technologies seem to provide important opportunities to increase the connectivity of innovation networks. He notes that productivity in research and learning activities can be enhanced providing networks remain open and not exclusive. While this may be partially true, I believe that there is considerable scope for testing this assertion. One of the starting points will be having a better understanding of how new communication technologies actually enhance or inhibit communication. Paradoxically, it is not evident that the new communication technologies automatically enhance communication, let alone that sort of communication which is important for innovation in firms.[6] Vital informal information networks could be threatened or distorted by networked office technologies leading to less effective information flows. Finally, there is no guarantee that openness will be preserved on the emerging communication networks. Exclusivity rather than inclusion is a business feature of many of these networks and this may have a detrimental effect on innovation.[7]

Antonelli's observations provide a basis for further research. I have suggested three possible areas for further investigation ranging from the large scale of questioning how countries copy technology policies from each other and the wisdom of doing that through to determining how communication technology may mediate communication at the level of the firm. In all these research areas, a better understanding of the role of information and its flow will be necessary.

Notes

1. See for example: R. Nelson (ed.), 1993, *National Systems of Innovation* (Oxford University Press, Oxford); H. Irwin and E. More, 1991, Technology transfer and communication: lessons from Silicon Valley, Route 128, Carolina's Research Triangle and Hi-tech Texas, *Journal of Information Science* 17:273-280; C. Maguire, E. Kazlauskas and A. Weir, 1994, *Information Services for lnnovative Organisations* (Academic Press, San Diego, CA); and S. Macdonald and C. Williams, 1992, The informal information network in an age of advanced telecommunications, *Human Systems Management*, 11:77-87.
2. D. Massey, P. Quintas and D. Wield, 1992, *High Tech Fantasies — Science Parks in Society, Science and Space* (Routledge, London).
3. J. Braithwaite, 1994, A sociology of modeling and the politics of empowerment, *British Journal of Sociology* 45:445-479.
4. N. Rosenberg, 1994, *Exploring the Black Box — Technology, Economics and History* (Cambridge University Press, Cambridge, UK).
5. W.E. Bijker, T.P. Hughes and T. Pinch, 1989, *The Social Construction of Technological Systems* (MIT Press, Cambridge, MA).
6. R. Van Dierdonck, K. Debackere and M.A. Rappa, 1991, An assessment of science parks: towards a better understanding of their role in the diffusion of technological knowledge, *R&D Management* 21(2):109-123.
7. S. Macdonald, 1992, "Information networks and the exchange of information", in C. Antonelli (ed.), *The Economics of Information Networks* (North-Holland, Amsterdam), 51-70.

CHAPTER 4

Comment: David Cracknell

I doubt whether, in practice, many firms know whether their R&D is truly a profit maximizing activity. They are more likely to be trying to reduce costs, extract competitive advantage or innovate products for particular market segments, all of which should have positive impacts on profits.

I am surprised that there is no discussion concerning patents and intellectual property rights. Some firms (particularly drug companies) rely on creating "monopolies" in particular market niches to justify their R&D.

It would also be interesting to explore the relationship between the R&D function and other aspects of company activity, as the BT experience would suggest to me that it is very important in setting the research agenda. In the last 10 years we have moved away from the notion that the scientists decided what to do. Most R&D now needs to attract a sponsor from the marketing, product management or operational departments, and many planned projects that cannot attract such sponsorship are dropped.

CHAPTER 5

Technological change

G.M. Peter Swann

1. Introduction

This does not pretend to be a comprehensive literature review. My aim in writing this chapter is to identify what are, from my perspective, some of the most important aspects of the current research agenda in the economics of technological change. A more comprehensive review of the most important items on the technological change agenda is in the recent book edited by Granstrand (1994).

For the most part, I shall have less to say about the social or management research agenda — not because I am uninterested in these, nor because I have no knowledge of them. Rather, my aim is to set out what I think economists should be doing in this area. Part of my message is that economists should be learning from what management and social scholars already know — and I daresay some knowledge transfer in the opposite direction could be advantageous too.

In this essay I want to discuss five broad questions about technological change. How should it be categorized and measured? What is needed to make it happen? What is the demand for it? What does it do to economies and societies? What (if anything) should be done to control it? The aim in each case is to briefly review some of what we know, and then list the issues that should, in my opinion, be receiving attention in future research.

These are not of course new questions, nor are they exhaustive. I think I have something new to say about each one, though I shan't pretend to survey all the other things that are already known. Moreover, this is an idiosyncratic list, but I make no apology for that. Finally, the discussion here will make frequent reference to issues raised in other chapters of the book.

2. How should we categorize and measure technological change?

At a microeconomic level, certainly, economics is a pretty imprecise taxonomic science. Compare it with botany, say. In botany, there is a long tradition of classifying plants in terms of some detailed measurements, and then putting them into families. New (innovative) hybrid rhododendrons, for example, can be placed into appropriate classes (Cox and Cox, 1988). In general — forgive the pun — a tree structure works for classification here. New categories can be identified as developments of old, though in principle categories may join together again. There is obviously a fractal character to plant innovation — and I use the term innovation advisedly since much of the economically interesting hybridization in rhododendrons (at least) has been man-made rather than naturally occurring.

In economics, by contrast, classification and taxonomy for product innovation is rather poor. We still measure innovative activity, for the most part, by patent counts or R&D expenditure though a growing programme of making direct innovation counts is under way (Archibugi *et al.*, 1991) following on from the classic SPRU innovation database (Townsend *et al.*, 1981).

It would be tedious to revisit at length the debate about the merits and demerits of patent counts. For many mapping purposes they are the only way the questions can be approached at all, and their use by Patel and Pavitt (1992), Cantwell (1994) and Archibugi and Pianta (1992) shows that many interesting things can be learnt using them. There is no other way (at present) that large scale and detailed cross-industry and cross-country comparisons can be made. But at the same time, we have to recognize that they do have problems as a measure of innovative activity: first because they measure invention rather than innovation; second because the value of a patent has a very high variance, with most worth little, but a few worth an immense amount; third they are not the only method of appropriation; and hence fourth, the use of patents varies from industry to industry.

Again, no one would seriously argue that R&D data are no use to them. Such data have many advantages: first because as a monetary measure R&D can be compared with other investment data; second because it takes account of cross-industry variations in patenting propensities. But using the Pavitt (1984) industry categories, we know that not all industries depend on R&D for innovation — and indeed Pavitt *et al.* (1987) showed that the ratio of innovations to R&D spent is high in small firms — who can produce innovations using little (recorded) R&D data. There are other reservations of course: R&D is an input not an output. And Vickery has shown that R&D may be a minority share of total innovation cost (cited in OECD, 1992).

When we use imperfect count data we run into what I have termed the "mixed bags" problem (Swann, 1994). The use of innovation counts is to me a welcome step forward, though some have suggested that these are just as much a mixed bag. And to take up again our comparison with botany. Botanists do count species, but they don't only count. They also measure the extent to which one hybrid (innovation) differs from the pack. They describe and map the evolution of hybrids.

This brings me on to a long standing item on my research agenda. I think it is essential in due course for economists of technological change to get involved in detailed mapping of product innovation and technological change. Some very interesting studies are by Trajtenberg (1990), Saviotti and Trickett (1992), Payson (1994), and this author has also done some work in this direction (Swann, 1985, 1986; Swann and Gill, 1993). It is immensely labour intensive, and a rather thankless task — in that there are many more effective ways of generating "top-journal" publications. On the other hand, my feeling is that the social value of this activity is quite large.

Consider the old distinction between process innovation and product innovations. When the first is "pure" — that is to say the process changes but the end product is unchanged — then as far as the end product is concerned, it is usually possible to map the process innovation into a cost or price reduction. Economists are pretty good at mapping price reductions — and even better at mapping price increases!

But now let us turn to the product innovation. Some writers would further sub-divide this into three categories of product innovation: (a) those that represent an improvement in one or more existing characteristics but no new features; (b) those that represent an addition of one (or a few) new features or characteristics; and (c) those that represent a completely new product. It could be argued that (c) is a limiting case of (b) in that if an innovation represents the addition of a very large number of new characteristics, then the characteristics space required to represent the innovation is so different from that required to represent existing products that they are effectively different products.

Economic theory is tolerably good at analysing (a) — preferably in two dimensions at most — and empirical studies have been quite good at mapping change in characteristics scores — again, so long as not too many dimensions are involved. Theory has had much less to say about (b) — generally it requires a simulation approach, and despite the interest of the question, many theorists do not find simulation an attractive methodology. Empirical mapping of (b) has been limited too. It is the stuff with which technological historians deal, but few economic studies that I have ever seen systematically map out the expansion of characteristics space. [This author has contributed one theoretical and one empirical study in this area: Swann, 1990; Swann and Taghavi, 1992, Chapter 17.]

Perversely, economic theory is much better at analysing (c). Actually, this isn't so surprising given the perception of some other social scientists that utility theory is essentially empty, and so can be "filled". In the Arrow-Debreu theory (Debreu, 1959), if product A and product B are not identical, then they are treated as if they are completely different products. The theory offers absolutely minimal predictions about how the demand for A and B will be related — the only predictions of consumer theory are, adding up, homogeneity, symmetry and negativity. The only parameters that describe the direct connections between demand for A and B are elasticities of substitution and cross-price elasticities of demand — and these are wholly empirical. For the traditional Arrow Debreu theory, A and B could be chalk and cheese or two different types of cheddar cheese. Empirically, however, it doesn't seem to me that economists have gone far in distinguishing (c) from (b) or even from (a).

There are of course some difficult conceptual problems here. How do you know if you have a new characteristic? Those using consumer report data or trade press comparisons could simply reply that a new product A has a new characteristic if that new characteristic has to be added to an existing comparative table in order to do justice to product A. But characteristics as measured in the trade press are sometimes answers to a question: does product A do this? If yes, then it has the characteristic; if no, then it doesn't. Until the question is posed, there is (in a data sense) no such characteristic. Bacharach (1991) talks about the interrelations between natural language, product description and demand. If the question is posed and the answer is yes, for at least one product, then the characteristic exists. The posing of the question, itself, is a social rather than a technological issue: we know from ethnographic and sociological studies that the uses to which a technology is put are underdetermined by the technology itself, and that many technologies find unexpected uses thanks to users making experimental and unscheduled uses of the technology. The dimensions of characteristics space could be (in part at least) a sociological issue.

There is little economic theory to inform this, and very little measurement either. Perhaps the greatest success is the literature on the hedonic technique and quality adjustment of price indices. One of the outstanding exponents of this technique, Jack Triplett, has however written a pessimistic "intellectual autopsy" of the technique (Triplett, 1990).

Another common distinction runs into equal difficulties, that between incremental and radical change. It seems very hard to distinguish between the two *ex ante* — and even *ex post*. The reason for this is perhaps that it is not just the size of the step in characteristics space that determines whether a change is incremental or radical. Large incremental steps are still incremental while some small but unexpected changes can be radical. Tushman and Anderson (1986) pointed out that whether change appears radical or simply incremental depends on whether it is compatible with existing competencies (i.e. competence enhancing or at least competence sustaining) or incompatible (and hence competence destroying). The issue is therefore not so much the size of the step, or indeed whether it traverses several new characteristics, but whether existing competencies were able to deal with this change. That is very hard to say *ex ante*: it requires not just a multidimensional map of the technology and its advance, which is hard to construct, but also an equally multidimensional map of competencies, which is even harder — though see Patel and Pavitt (1994).

And we know from Schumpeter that "competition from the new commodity, the new technology, the new source of supply, the new type of organisation ... is as much more effective than the other (price competition) as a bombardment is in comparison with forcing a door ... that it becomes a matter of comparative indifference whether competition in the ordinary sense functions more or less promptly" (Schumpeter, 1954:84-85). If we have progressed little in mapping product innovations, we have gone even less far in mapping organizational innovations, architectural innovation (Henderson and Clark, 1990), networking innovations (see the chapter by Cristiano Antonelli), network firms, and network products. Yes, mapping these innovations is much harder than in botany, where things are relatively stable. But all the more reason to invest in it.

This to my mind maps out a huge agenda for the economics of technological change. After all it seems perverse in the extreme to a Schumpeterian that we put so much effort (relatively speaking) into measuring rates of price inflation, when we have got almost nowhere with measuring the altogether more important phenomenon of product innovation. It seems absurd to quote percentage rates of inflation to 1 decimal place (e.g. 2.4%), when we don't know even to one significant figure the rate of product improvement.

3. What is needed to make technological change happen?

I can't give an exhaustive account here — who could? I shall pick on three themes that interest me especially, and to which I think economists can contribute something. In other important areas, I suspect it is the management and social scholar who will have more to say than the economist.

The three questions are these. What sort of market structure is needed to encourage innovation? When are networks essential for innovation? And when is IP protection essential to support innovation?

3.1 Market structure and innovation

In much of the economics literature, we are just as interested in the effects of innovation on market structure — and I return to that below. For now, I look at the other side of the relationship: which market structures are most conducive to innovation?

The "folklore" of the subject suggests an inverse-U-shaped relationship between concentration and innovation. The unthreatened monopolist would find it easy to finance innovation, but has little incentive — since he has 100 per cent market share anyway. In perfect competition, by contrast, there is plenty of motive, but no means and no appropriability. As George Bernard Shaw observed in Man and Superman of the marriage duopoly: "(it) is popular because it combines the maximum of temptation with the maximum of opportunity". Here the folklore is less precise: it may be duopoly or it may be a more fragmented oligopoly which combines the maximum of incentive with the greatest means. Nonetheless the folklore has a theoretical rationale and some empirical support. But this empirical support tends to derive from studies that measure innovative activity by means of patent counts or R&D. To the extent that the competition end of the spectrum identifies markets with many small firms, who may be innovative but do not do R&D and do not patent, then this inverse-U may be something of an illusion (Pavitt *et al.*, 1987).

I don't say that this seam will yield no more gems, but it does look as though some of the most interesting potential findings about the effects of market structure on innovation will come from digging elsewhere. In particular, I suggest the most interesting findings may come from thinking about the dynamics rather than the statics of market structure.

As some have stressed (Futia, 1980; Hammond, 1984) Schumpeter argued that it is not so much the static oligopoly market structure that promotes innovation but the prospect of a temporary and local monopoly. Such prospects may be high in a reasonably concentrated oligopoly, but this is not the only setting in which such prospects arise. Entrants with a radical innovation may be well placed to benefit. Are those startups that survive disproportionately successful at radical innovations? While some of the empirical literature on innovative MNCs suggest that this is romantic nonsense, the organizational literature on organic form and competence-destroying technical change would suggest that they might be. And if so, what view should we take of market share turbulence?

It has been identified by many writers (for example, Davies *et al.*, 1991; Audretsch, 1992; Swann and Gill, 1993) that a stable concentration may be accompanied by various degrees of market share turbulence — meaning that the identity of the top four may change even if the share of the (changing) top four is constant.

This is undoubtedly a consequence of the rate and type of innovation. But does the rate of turbulence also condition the ability of an economy to deliver technological change? Some would say yes. Schumpeter's creative destruction makes this quite explicit: the innovative process may destroy, but the value of what it creates exceeds the value of what it destroys. The remarkable, and still not widely known work of Downie (1958) points in the same direction. Business cycle optimists have suggested that the "silver lining" in business cycles is that they encourage an economy to restructure so that the inefficient are selected out while the efficient survive. Many evolutionary economists might however query whether the selection mechanism implicit in the UK economy, for example, is optimized to select the long-term innovators or rather those with a view to the short- to medium-term. And there is very little empirical evidence and detailed theoretical analysis of this question. The analysis of hysteresis in macroeconomics (Cross, 1988) tends to argue in the opposite direction: that the business cycle destroys investments, and hence is damaging to long term growth.

Market share turbulence does the same, but it also strengthens the prospects of the new, and may therefore promote long term growth. Or it may not — again, we have little theory or evidence on this. Not indeed that the long term growth necessarily compensates for the dislocation caused by cycles — but at least this gives us a new view on the role of the business cycle.

3.2 Networks and innovation

Moving away from the simple (or not so simple) relationship between market structure and innovation, we should ask what is the effect more broadly of industrial structure on innovation. We shouldn't just ask what is the implication of the distribution of firm sizes (as well as vintages and competencies) for innovation, but also what is the effect on innovation of the relationships between firms. Given market structure (in terms of the distribution of firm sizes) how does the degree and form of networking amongst them influence the ease and extent of innovation? If firms generate potential spillovers from their innovative activities that could benefit others, then we can envisage circumstances in which networking enhances innovation. Does geographical proximity matter in this networking? One view here is, yes, where the technological knowledge to be transferred is tacit, but no if it is codified. Some of the literature on clusters builds on this distinction between tacit and codified knowledge. And certainly this literature is right to ask — even if an answer is difficult — how rapidly existing knowledge is becoming codified, and if so, whether the speed of codification exceeds the rate of generation of new tacit knowledge.

One of the themes emerging from this literature is that if the flow of knowledge around a network is constrained, then so also is the growth of knowledge and the rate of innovation. To promote this information flow calls for open networks and that in turn calls for standards of communication — whether technical or informational (David and Greenstein, 1990). Indeed, this observation opens up a very important area of tension between the need for networking in the innovative process and the extent to which intellectual property protection can constrain that networking. We return to this shortly.

The chapter by Cristiano Antonelli has already picked up many of the reasons why networks are essential for innovation, and I have little more to add to that.

3.3 Intellectual property protection and innovation

The traditional economic rationale for patenting was as a corrective to the market failure that could arise from R&D spillovers. If investment in R&D by company A generates knowledge that spills over (at no cost) to company A's rivals, then company A does not appropriate the full benefits of its investment, and this can lead to underinvestment. In essence the social value exceeds the private return, and hence at the margin some socially profitable projects will not be privately funded because the cost exceeds the private return even if the social return exceeds the cost. Patenting, and other devices to enforce property rights can internalize some of these spillovers (through royalties), or compensate by offering a temporary monopoly. If the patent scope and length are adjusted optimally, so the argument goes, the private returns can be boosted to correct this market failure. So far so good. But note that this is not an unlimited monopoly right: it simply says that the expected rent from the patent should bring the private value of the investment into line with the discounted social value of the investment.

I set aside here the important debate about whether spillovers are really an aspect of market failure, and concentrate instead on whether IP protection is in practice set according to this basic economic principle. I would argue that in some areas of new technology (e.g. software, networks and biotechnology) intellectual property protection may possibly be too strong — at least when viewed from an economist's perspective. By that I mean that the costs of allowing IP protection at its present level could outweigh the benefits. Some would argue that compulsory licensing attenuates any efficiency loss from over zealous IP protection, and simply leads to a transfer payment. I would agree that compulsory licensing helps, up to a point, but it still doesn't overcome the efficiency objections.

The argument is familiar enough. Setting aside the dynamic efficiency issue, IP protection that denies other use of the knowledge (at any price) is clearly inefficient in a static sense. The social cost of the pure spillover is effectively zero. This is knowledge that may even be in the private domain (post patenting), and while there may be costs of assimilation and comprehension, there is no appreciable cost of spillover. In monopoly terms this is a deadweight loss, and in the context of some biotechnology patents, a potentially very large loss too. Compulsory licensing corrects some of the deadweight loss — in that at least some of those willing to pay the royalty are able to use the knowledge. But if the royalty is set at the monopoly price it doesn't correct the deadweight loss corresponding to those who are closed out of the market but value the knowledge above the marginal cost of dissemination (nil).

This is not perhaps wholly fair to patents. While they restrict the use of knowledge, they do not restrict the dissemination of knowledge. In that sense they are arguably preferable to commercial secrecy as a means of protecting intellectual property. But as Don Lamberton pointed out at the Canberra Workshop, many patent searches are made simply to check that an independent discovery does not infringe some pre-existing property right — and not to learn new knowledge. In terms of the pure economic logic, if A patents a discovery first but B makes a subsequent independent discovery, then there has been no spillover from A to B, no market failure to correct, and no royalty should be collected. The social value of A's finding is equal to the private value.

Moreover, a comparatively recent perspective on R&D and spillovers has caused us to question whether spillovers really do undermine the incentive to do R&D. If as Cohen and Levinthal (1989) argue, much R&D is learning rather than invention, this R&D is necessary because it is costly to assimilate spillovers. In that case, as Geroski (1994) points out, it is not clear that spillovers will reduce the incentive to do R&D; on the contrary, the existence of spillovers from another firm can increase the incentive to undertake learning R&D. This argument needs to be clarified a little. If A's R&D generates a spillover to B, then even if B needs to spend on learning R&D to assimilate the spillover, it still is a spillover. To that extent the private return to A is less than the social return on the investment. But the argument is that in the presence of the spillover from A to B, company B finds it profitable to spend an additional amount on learning R&D to assimilate the spillover. This additional R&D may partially offset the original spillover if A benefits from B's investment in R&D for learning. If this last phenomenon happens, then certainly the distance between the private return to A from his R&D investment and the social value of that R&D is reduced.

It seems a bit surprising at a time of rapid and accelerating rates of technological change to find some observers arguing that innovative activities are being constrained in an unsatisfactory way because of inadequate appropriability. Some would argue, of course, that it is not so much the absolute level of IP protection that one should be concerned with, but rather that one must recognize that differences in the level of IP protection in different countries is distorting.

We return to the tension between IP protection and networking. Macdonald (1990) has shown how US export restraints in high technology areas could in fact rebound by constraining information flows around the world network. Shurmer and Lea (forthcoming) have shown how IP protection can make for serious difficulties in the attempt to reach IT standards. If a vital component is subject to tight IP protection and the owner is unwilling to license, then any institutional standard has in effect to design around that component. This can render the institutional standard unsatisfactory, and can lead to the imposition of a proprietary solution.

I won't labour this any further. Suffice it to say that IP protection lies in tension with networking, and I submit that in a variety of cases, the promotion of networking may be more important to the promotion of innovation than the market failure attributable to insufficient IP protection. But at any rate, this is an important empirical and conceptual research agenda for further study.

4. What is the demand for new technologies?

There are two questions I wish to address here. First, what new techniques are needed to estimate the demand for new technologies? Second, do high market sales imply a strong underlying demand for technological change, or is there a sense in which market sales can actually overstate the value derived from technological change?

4.1 New techniques

Standard demand analysis can only really help us for forecasting how markets will behave in response to prices and qualities that are part of past experience. When we talk of advances in the technological frontier, and low price levels never before seen, then we are into the realm of extrapolation — as far as traditional price-quantity demand analysis is concerned. And extrapolation, as is well known, is a risky business.

As we noted above, the Arrow-Debreu model treats new products as completely different products, so to estimate demand for these new products requires a time series of data on prices and sales. But by definition we don't have that for new products.

If we treat new products as a better version of an existing product, and assume that demand for these new products can be modelled using quality-adjusted prices, then we can at least using an existing demand model to project demand at these new quality-adjusted prices. But again, this is almost certain to require extrapolation outside the range of past history.

In any case, quality-adjusting as a procedure doesn't really capture the significance of quality enhancing technological change. To take one obvious example in microelectronics. The continuing miniaturization of electronic components led to lower prices per bit of memory, but that is only a small part of the overall pattern. Miniaturization also led to application of electronics in all kinds of setting where it would not have been possible before: watches, cars, washing machines, and so on. To analyse the demand for new electronic components by quality-adjusting a price index would seriously miss the point.

All of which implies that to make progress in modelling the demand for new products, we need to work in a product characteristics context. And certainly, the advances in modelling the demand for product quality have been significant, notably the use of logit and probit modelling in a discrete choice setting (McFadden, 1973; Hausman and Wise, 1978) and other continuous approaches to modelling quality demand (these are summarized in Swann, 1986). Again, of course, quality improvements are likely to lead us into the realm of extrapolation — how can this be avoided? And indeed, that is probably the least of our problems. For if, as indicated above, some of the technological change introduces new product characteristics, then we again face the problem that we are breaking out of the space with which we are familiar. Modelling the demand for a new characteristic generates the same difficulties as modelling the demand for new products — namely that there is no history of demand to draw on.

From my perspective, this is still a major research agenda. We can't avoid the extrapolation problem — after all it would hardly be technological advance if we could. But what we can hope to do is find a way of mapping technology characteristics into a smaller set of fundamental characteristics — so that the addition of a new technology characteristic does not necessarily expand the space of fundamental characteristics. This is reminiscent of the household production model (Becker, 1965) in which consumers buy products not so much for the characteristics they offer, but for the household services that can be generated by combining these characteristics with home labour. The problem with household production theory has always been that if it is difficult to estimate a demand structure over actual characteristics, then it is even harder when we postulate a set of latent variables (i.e. fundamental characteristics) which have themselves to be estimated. This always made it hard to put a production structure on the household production model. But that does seem to be an approach we must take if we are to analyze the demand for such technical change in a stable space of fundamental characteristics.

Another approach which is used widely as a method of pragmatic if "informal" demand projection is to assemble a list of discrete activities that happen at present, and elaborate what would have to happen to costs and performance of the new technology for it to be profitable to perform these activities using new technologies. A number of early evaluative studies of microelectronics (Barron and Curnow, 1979; Evans, 1979; Bessant *et al.*, 1981) set out such lists, and commented informally at least on what would have to happen to the technology before it would be used in that context.

This is fine for substitution: technology X doing function A that was previously performed using technology Y. It is also fine for enhancements of existing activities. But what about new demand arising from new applications? The sociology of IT use has stressed that the ultimate uses of a technology may extend some way beyond the immediate uses for which it was developed. While we make fun of some early computer executives for thinking that a few machines could satisfy world computing needs, how could we have predicted the diversity of applications for which they are now used? This sort of foresight is not only difficult; in some sense it may be impossible if we accept the proposition that it is impossible to forecast innovative applications of a technology.

4.2 Market sales but not user value?

Do strong market sales for the latest upgrade to an established technology imply that users attach a high value to such technological advance? This may appear to be an absurd question: if users pay for something they must value what it offers. But suppose that demand increases because of what we could call pure epidemic diffusion, and at the same time an improved quality of product is marketed at no extra price. Demand increases, the higher quality product is purchased, and it looks as if there is a causal connection. But in fact, the typical consumer need only attach a very small marginal valuation to quality to choose the higher quality in preference to the older product. The strong growth in demand does not imply that the consumer attaches a high marginal value to technological change.

This ties in with the competitive pressure models of diffusion (for example, Reinganum, 1981) which suggest that the rate of diffusion of a new technology is determined not so much by the spread of information or by increased demand as prices fall (or quality rises) but by the fact that company A's adoption of the technology increases the competitive pressure on B to adopt. There is an element of prisoner's dilemma here. If duopolists A and B both adopt they incur additional costs but neither ends up ahead; if they could have agreed not to adopt they would have saved themselves the cost and their relative competitive positions are the same. In this setting strong demand does not necessarily imply high social value from adoption.

This model of demand generated by competitive pressure may seem unduly pessimistic. It may also seem irrelevant to consumer (rather than institutional) demand, but a little reflection shows that it is not. Bourdieu's theory of distinction of taste (Bourdieu, 1984) implies in economists' language that it is the prospect for group A to differentiate themselves from another group B by consumption of X that generates the market demand for X. If B also consume X, then A gain no benefit from it. It is difficult to say how important this is, but many are of the view that we have a serious need for an evolutionary theory of demand, and this seems a good place to start.

The agenda here overlaps considerably with that set out in Chapter 3. But from my point of view, the key to modelling demand for new technologies lies with the ability to identify how to plot these innovations in an ultimate characteristics space, and to construct a serious evolutionary theory of demand.

5. What are the effects of technological change?

Some of the effects are discussed in other chapters [6, 7, 9 and 10]. Here I want to raise four important agenda items: ICTs and unemployment; ICT and regional development; Dynamics of technological change and market structure; Benign and not-so-benign effects of technological change.

5.1 ICTs and unemployment

At a fascinating recent lecture, Chris Freeman (1994) argued that at a global level the net employment effects of new technologies depend on the balance between the jobs displaced by labour replacing technological change and the new jobs created to produce and service new technologies and markets. On a national scale, the same calculus applies, though it is necessary to add into the equation the net national gain or loss of jobs in existing industries resulting from the geographical relocation of activity. For South East Asian countries there was probably net job creation, in part because they were the beneficiaries of relocation of production, but in many developed countries there was a net loss of jobs.

The issue for developed countries is whether new technologies can generate new markets and new jobs fast enough to compensate for the loss of jobs from labour replacement and the loss of jobs from production relocation to S.E. Asia. We know from some case studies that the simplistic model of labour replacing investments in ICT has been misleading because the use of ICT has in fact been coupled by product or service enhancements. Nevertheless, these sources of demand for labour are not enough on their own to offset the other source of job destruction.

I would argue that we really need some theoretical developments here. To my mind we don't have an adequate theory of how new technologies create new demands for labour — perhaps indeed because of our limited understanding of the demand for new products (see above).

5.2 ICTs and regional development

This issue is addressed in chapter 10, so I shall only make two observations here. Some technological optimists would argue that the widespread availability of ICT networks, and widespread access, can help to correct regional imbalances in economic development. To put it in simple language, the availability of (for example) an online catalogue on the ICT network can help to increase the competitiveness of the village shop in the eyes of city dwellers. I have a problem with this argument. It is fine as far as it goes, but it doesn't look at the reverse effect. Yes the online catalogue makes the village shop more competitive in the city, but it also makes the city department store more competitive in the village. It is not clear which effect will dominate. Moving to the transport context, some episodes of transport history suggest that the optimism that rail networks would disperse industrial activity were in fact misplaced, and the net result has been further concentration.

This is an empirical matter which has been studied by a number of economic geographers (for example the CURDS centre at the University of Newcastle, UK). But we need more evidence, and we also need further theoretical work here.

A related point is whether ICT will act as a substitute for transport, and hence lead to less transport — and hence less fuel consumption and air pollution. The optimists would point to the simple comparative static substitution effect and say that where ICT is a substitute for transport or travel then this desirable effect is achieved. There are two potential problems with this. First, what about the income effect? The replacement of costly transport by cheap ICT transmissions may lead to an expansion in trade — and hence some new demand for transport. This income effect could be less than or greater than the substitution effect. Second, are transport and ICT transmissions always substitutes? Sometimes, surely, they are complements. New long distance trade activities can require a combination of ICT transmissions and transport to service them. To the extent that new global ICT networks expand long distance trade, then long distance travel can increase too.

The main agenda item here is the same in each case. The net and long term effects of ICT networks on the regional dispersion of industrial activity and transport may operate in the opposite direction to what is expected on simple intuitive grounds. Whether it does or not is an empirical question. But in addition, it is important that we develop a fuller theoretical framework for analysing this question.

5.3 Dynamics of technological change and market structure

We commented above on the effects of changing market structure on innovation; here we comment briefly on the reverse effect. Much energy has gone into analysing the question: is rapid technological change concentrating or deconcentrating? If we believe that technological change is deconcentrating then we tend to consider that a good thing — given our general preference for competition? I don't have much to add here that is new. But what about the dynamic effects of technological change on market structure? To what extent does rapid technological change lead to market share turbulence? Is turbulence good, bad or indifferent? These are questions about which less is known, and they would go on my research agenda.

5.4 Benign and not-so-benign effects of technological change

It is hardly a new observation that technological change can be good for some and not so good for others. The Luddite attitude to technological change was not necessarily born of ignorance but a correct economic analysis: in the short term — at the very least — labour replacing technological change is not good for the labour made unemployed by that change, unless replacement employment is provided.

Given that there are clearly a mix of benefits and costs from technological change, then even if the former outweighs the latter (and we trust that it does), we still need an account of each. In fact economics has made remarkably little progress in analysing this question — still less in quantifying it.

We suggest a simple five way classification of technological change according to the spillover effects of such change on third parties who do not finance the technological change themselves.

Classification	Magnitude of Spillovers
Benign	Positive
Private	None
Redistributive but Benign in Aggregate	Benefits to adopters outweigh negative externalities
Neutral (Purely Redistributive)	Benefits to adopters exactly offset by negative externalities
Not Benign	Negative externalities outweigh benefits to adopters

For any given innovation, it may be difficult to classify, but it is something we should try to do. Medical innovations (new antibiotics, for example) appear wholly benign at first sight because the effect on the adopter is beneficial and the health risks to others are attenuated — so there are beneficial spillovers. Some would take a more pessimistic line and suggest that the existence of antibiotics may promote more virulent mutations that will be harmful to later generations. I don't have the medical competence to comment on this.

Innovation in the social domain is rarely a private matter — almost by definition — but an example would be the adoption of an (own) labour saving innovation by a single person, who spends his additional leisure reading a book instead. Even then some spillovers might occur from his new-found erudition.

Innovations that are redistributive, but benign in aggregate, would include process innovations that cause some disruption, possibly some job dislocation, but ensure the survival of the organization, or raise national income. These are of course potentially Pareto-improving innovations, but as we know well from basic welfare economics there will only be an actual Pareto improvement if those dislocated are adequately compensated.

Innovations that are purely redistributive are those for which a zero-sum game operates. They could include trivial innovations which redistribute market share from one company to another, and have a negligible effect on consumer welfare. Some second order marketing and promotional innovations may be of this sort.

Finally, the non-benign innovation is unfortunately not simply a theoretical construct. It would perhaps be invidious to pick on any one technology, and in any case whether or not a technological innovation is malignant will depend not simply on the technology itself but also on the manner and the environment in which it is used. Some would place the use of loud personal stereos on crowded trains in this category. The pleasure to the one willing listener is outweighed by the irritation to the many unwilling listeners. On the other hand the quiet use of personal stereos in a private place is a private matter. More seriously, the abuse of some computer innovations can threaten very basic rights such as security and integrity of personal data. While such innovations may be quite benign in the right hands, or the right competitive setting, they may not be benign in the most competitive setting. Gresham's Law rears its ugly head again: those who abuse the innovation may be able to undercut or outcompete those who play fair.

6. What (if anything) should be done to control technological change?

Martin Cave's chapter on Regulation (8) addresses several aspects of this question. I have only a little to add.

Using the classification of section 5.4 above, it seems clear that there is a *prima facie* need for legislation to prevent the non-benign use of technology. Legislation on computers and privacy, on noise pollution (and other sorts of pollution), on the use of parts of the radio spectrum, on the use of powerful new drugs, and so on, are examples of what is done to control this sort of technological change.

Should we also seek to control the use of purely redistributive technological change? I would argue yes — where practicable. Such innovation gives innovation a bad name. One example is the over-rapid introduction of new models of high technology products, trivially differentiated from each other. There would sometimes appear to be a case for limiting the rate of new product introductions, even if this could be construed as collusion to keep down R&D costs. Indeed, we know from the literature on product proliferation (Scherer,

1979; Schmalensee, 1978) that proliferation of minimally differentiated products can have a variety of costs to the consumer which can outweigh the diminishing additional utility arising from greater choice. Product proliferation over time can likewise be seen as a barrier to entry.

On the other hand, it would be a serious mistake to seek to control all innovations that are redistributive, but clearly benign in aggregate. What is required in such cases is serious attention to compensation of the losers. If ICTs are in this category, then it seems appropriate to allow the use of such technologies to proceed, but with job creation schemes to reemploy those made unemployed. The "leave it to the market" argument does not address this.

For me there is a very large policy agenda here, which will draw on the measurements made in the previous section. It has to be recognized that some technologies in some competitive settings can have significant negative externalities, and we have to think about ways in which to identify these when they happen, and devise policy instruments to control or at least to compensate.

References

Audretsch, D., 1992, The technological regime and market evolution: the new learning, *Economics of Innovation and New Technology* 2(1):27-36.

Archibugi, D., S. Cesaratto and G. Sirilli, 1991, Sources of innovative activities and industrial organization in Italy, *Research Policy* 20:299-313.

Archibugi, D. and M. Pianta, 1992, *The Technological Specialization of Advanced Countries* (Kluwer Academic Publishers).

Bacharach, M., 1991, Commodities, language and desire, *Journal of Philosophy* 87(7):346-386.

Barron, I. and R. Curnow, 1979, *The Future with Microelectronics* (Frances Pinter Publishers, London).

Becker, G., 1965, A theory of the allocation of time, *Economic Journal* 75:493-517.

Bessant, J.R., J.A.E. Bowen, K.E. Dickson and J. Marsh, 1981, *The Impact of Microelectronics: A Review of the Literature* (Frances Pinter Publishers, London).

Bourdieu, P., 1984, *Distinction: A Social Critique of the Judgement of Taste*, transl. R. Nice (Harvard University Press).

Cantwell, J., 1994, *Transnational Corporations and Innovatory Activities* (Routledge, London).

Cohen, W.M. and D. Levinthal, 1989, Innovation and learning: the two faces of R&D, *Economic Journal* 99:569-596.

Cox, P.A. and K.N.E Cox, 1988, *Encyclopedia of Rhododendron Hybrids* (B.T. Batsford Ltd, London)

Cross, R., 1988, *Unemployment, Hysteresis and the Natural Rate Hypothesis* (Basil Blackwell, Oxford).

David, P.A. and S. Greenstein, 1990, The economics of compatibility standards: an introduction to recent research, *Economics of Innovation and New Technology* 1(1/2):3-41.

Davies, S., P.A. Geroski, M. Lund and A. Vlassopoulos, 1991, *The Dynamics of Narket Leadership in the UK Manufacturing Industry, 1979-1986*. Centre for Business Strategy Report, London Business School.

Debreu, G., 1959, *Theory of Value* (Yale University Press, New Haven).

Downie, J., 1958, *The Competitive Process* (Duckworth, London).

Evans, C., 1979, *The Mighty Micro: Impact of the Computer Revolution* (Victor Gollancz Ltd, London).

Freeman, C., 1994, Unemployment and the diffusion of information technologies: the two-edged nature of technical change. PICT Policy Research Paper No. 32 (Programme on Information and Communication Technologies, Economic and Social Research Council).

Futia, C.A., 1980, Schumpeterian competition, *Quarterly Journal of Economics* 94, 675-695.

Geroski, P.A., 1994, Do spillovers undermine the incentive to innovate?, in: *1994 Conference of Industry Economics: Papers and Proceedings Bureau of Industry Economics*, Australian National University.

Granstrand, O., 1994, *Economics of Technology* (North-Holland, Amsterdam).

Hammond, P.J., 1984, Must monopoly power accompany innovation, in: C. Seidl (ed.), *Lectures on Schumpeterian Economics* (Springer-Verlag, Berlin).

Hausman, J. and D. Wise, 1978, A conditional probit model for qualitative choice: discrete decisions recognising interdependence and heterogeneous preferences, *Econometrica* 46:403-426.

Henderson, R. and K. Clark, 1990, Architectural innovation, *Administrative Science Quarterly* 35:9-30.

Lancaster, K.J., 1971, *Consumer Demand: A New Approach* (Columbia University Press, New York).

Macdonald, S., 1990, *Strategic Export Controls: Hurting the East or Weakening the West?* (Economist Intelligence Unit, London).

McFadden, D., 1973, Conditional logit analysis of qualitative choice behaviour, in: P. Zarembka (ed.), *Frontiers in Econometrics* (Academic Press, New York).

OECD, 1992, *Technology and the Economy* (Organisation for Economic Cooperation and Development, Paris).

Patel, P. and K. Pavitt, 1992, The innovative performance of the world's largest firms: some new evidence, *Economics of Innovation and New Technology* 2(2):91-102.

Patel, P. and K. Pavitt, 1994, Technological competencies in the world's largest firms: characteristics, constraints and scope for managerial choice, STEEP Discussion Paper 13, Science Policy Research Unit, University of Sussex.

Pavitt, K., 1984, Sectoral patterns of technical change: towards a taxonomy and a theory, *Research Policy* 13:343-373.

Pavitt, K., M. Robson and J. Townsend, 1987, The size distribution of innovating firms in the UK: 1945-1983, *Journal of Industrial Economics* 35(3):297-316.

Payson, S., 1994, *Quality Measurement in Economics: New Perspectives on the Evolution of Goods and Services* (Edward Elgar, Aldershot).

Reinganum, J.F., 1981, On the diffusion of a new technology: a game theoretic approach, *Review of Economic Studies* 48:395-405.

Saviotti, P.P. and A. Trickett, 1992, The evolution of helicopter technology, *Economics of Innovation and New Technology* 2(2):111-130.

Scherer, F.M., 1979, The welfare economics of product variety: an application to the ready-to-eat cereals industry, *Journal of Industrial Economics* 28:113-134.

Schmalensee, R., 1978, Entry deterrence in the ready-to-eat breakfast cereals industry, *Bell Journal of Economics* 9:305-327.

Schumpeter, J.A., 1954, *Capitalism, Socialism and Democracy*, 4th Edition (Unwin University Books, London).

Shurmer, M. and G. Lea, forthcoming, Telecommunications standardization and intellectual property rights: a fundamental dilemma?, *ACM Standard View*, 2(4):forthcoming.

Swann, P., 1985, Product competition in microprocessors, *Journal of Industrial Economics* 33(1):33-54.

Swann, P., 1986, *Quality Innovation: An Economic Analysis of Rapid Improvements in Microelectronic Components* (Frances Pinter Publishers, London).

Swann, P., 1990, Product competition and the dimensions of product space, *International Journal of Industrial Organisation* 8(2):281-295.

Swann, P., 1994, Inference from mixed bags: the economic value of patent counts, innovation counts and the like. Centre for Business Strategy Working Paper 147, London Business School.

Swann, P. and J. Gill, 1993, *Corporate Vision and Rapid Technological Change: The Evolution of Market Structure* (Routledg, London).

Swann, P. and M. Taghavi, 1992, *Measuring Price and Quality Competitiveness: A Study of 18 British Product Markets* (Avebury, Aldershot).

Townsend, J., F. Henwood, G. Thomas, K. Pavitt and S. Wyatt, 1981, Innovations in Britain since 1945, Occasional Paper No. 16, Science Policy Research Unit, University of Sussex.

Trajtenberg, M., 1990, *Economic Analysis of Product Innovation: The Case of CT Scanners* (Harvard University Press, Cambridge, MA).

Triplett, J.E., 1990, Hedonic methods in statistical agency environments: an intellectual biopsy, in: E.R. Berndt and J.E. Triplett (eds), *Fifty Years of Economic Measurement: The Jubilee of the Conference on Research in Income and Wealth* (Chicago University Press, Chicago).

Tushman, M.L. and P. Anderson, 1986, Technological discontinuities and organisational environments, *Administrative Science Quarterly* 31(3):439-465.

CHAPTER 5

Comment: Meheroo Jussawalla

Technological change — particularly in the area of information services — represents the key engine fuelling economic growth and development in the global economy of the twenty-first century. Swann's chapter provides a concise and well-organized overview of some of the key research challenges facing economists in the field of innovation and technical progress. However, through the integration of additional information on the measurement of innovation, the development of New Growth Theory, and the deleterious face of technological progress, the piece can be strengthened.

First of all, as Swann correctly argues, the measurement of innovation remains a key problem for economists. Indeed, both patent counts and R&D data essentially measure the raw production of information rather than innovation. Patent counts may boom and R&D expenditures may surge; yet, technological progress may not necessarily result.

At the same time, the massive production of information associated with high patent counts and R&D growth is not always a positive economic force. The production of information is linked to the consumption of information, which has its costs. As Macdonald argues, "what has been ignored in the euphoria surrounding the declining production cost of information is the consumption cost of the information produced. Resources are required to use information, and the more information produced, the more resources are required."[1] These costs are particularly serious when they are accompanied merely with the production of *information* and not the production of *innovation.*

Additionally, R&D measurements run into a variety of difficulties. As Lamberton has added, "if R&D, as now conventionally measured, is taken as the expenditure on new knowledge creation, the net is too narrow in that this scientist's definition of R&D excludes a significant part of innovation costs that could properly be held to create knowledge. At the same time, much expenditure that is little more than the maintenance of stocks of knowledge and manpower will be included."[2]

Finally, it must be noted that patent counts and R&D figures say nothing about the relative efficiency of our efforts to gain technological progress.

Swann may wish to integrate these points into his call for economists to reexamine the ways in which they measure innovation.

Secondly, we can benefit from a closer examination of New Growth Theory. As the principal architect of New Growth Theory, Romer of the University of California at Berkeley and the Hoover Institute at Stanford University has moved a step beyond Solow's traditional mathematical growth theory, which views innovation as simply an exogenous variable and "a random occurrence".[3] According to Solow's model, technological innovation represents the key source of long term economic growth as measured in terms of output per worker.[4] However, Solow's model does not seek to explain the sources of technological progress. As a result, the model quantifies growth in total factor productivity indirectly. Using data on growth in output (Y), capital (K), and labour (L) and data on capital's share of output (_), the rate of technological change can be computed by first calculating the total contribution of capital and labour growth to changes in aggregate output. The portion of aggregate growth in output that cannot be attributed to either capital or labour expansion represents the growth in total factor productivity. This value, which is measured technical progress, is also known as the Solow residual.[5]

Building on the work of Solow, Romer has made significant strides in the ability of economic theory to assess the impact of innovation on growth and has expressed the relationship between technological progress and economic expansion in mathematical form. Furthermore, rather than simply accepting changes in factor productivity as a residual and exogenous phenomenon Romer has sought to understand the sources of technological progress by analysing the externalities linking the execution of certain economic activities to expansion of total factor productivity.[6] The inclusion of Romer's work in Swann's coverage would strengthen the analysis of recent scholarship in the field of technological innovation and economic growth. Additionally, an examination of Romer's work may serve to highlight several of the enduring questions and on-going debates which lie at the top of the research agenda in this field.

Thirdly, Swann's article remains rather optimistic about the impact of technological change on society. Indeed, we are operating in a global economy in which the mastery of technology represents the key determinant of economic growth. However, as Swann himself points out, many questions remain about the impact of technological innovation on economic development. For example, several publications in recent months have embraced Luddite analyses of technological innovation and have received a great deal of press coverage, even spawning a recent cover story in *The*

Economist. Indeed, unemployment can represent a negative aspect of technical change. As Swann points out, innovation is a double-edged sword, simultaneously giving rise to the use of technology as a substitute for the use of labour while also creating demand for labour in certain areas of the economy. However, beyond this, technological change can result in displacement and broad structural adjustments in the economy that can translate into a form of unemployment resulting from the stickiness of labour markets.

Furthermore, technological progress can also undermine the fabric of society in a variety of ways. First of all, the decentralization of technological know-how can threaten lives. For example, the increased availability of a variety of chemical compounds have made such terrorist acts such as the gas explosion in the Tokyo subway increasingly possible. As the popular film "Outbreak" pointed out, increasing rates of biological experimentation can accidentally lead to the spread of a deadly virus. Furthermore, there is a significant danger that technological advancement will outpace the need for careful ethical and moral deliberation about the use of such technology. For instance, the explosive growth of the field of genetic research suggests that society must begin to think about the potential harms of such practices as genetic experimentation and replication. Additionally, theorists from Sigmund Freud all the way to Alvin Toffler have written extensively about the deleterious impact that civilization-building, rapid technological advancement, and future shock can have upon the human psyche.[7]

Technological progress also has a profound effect upon the redistribution of economic resources both within an economy and between the industrializing and industrialized worlds.[8] A brief examination of Swann's citations reveals an exclusive European emphasis in his piece. This "Euro-bias" can be rectified through a greater examination of the impact of technological change on other developed and developing regions. Finally, technological progress can also affect government structures and the distribution of political resources in a society. Writing in the same vein as George Orwell did in his famous piece on "Politics and the English Language", Ian Reinecke has argued that technological progress has given rise to an oppressive regime of techno speak which has reduced "otherwise intelligent, articulate, knowledgeable people to silence on matters technological".[9] This silence has resulted in the exclusion of a large element of the population from the public debate on subjects of a technological nature, and, in the future, may threaten our very freedoms and system of government. "There have been too few protests by the non-technical at the tyranny of technological language",[10] Reinecke concludes. A greater emphasis on the continuing debate between the positive and negative aspects of technological innovation strengthens Swann's article.

Notes

1. Stuart Macdonald, 1983, Technology beyond machines. In: Don Lamberton, Stuart Macdonald and Tom Mandeville, eds, *The Trouble with Technology: Exploration in the Process of Technological Change* (Frances Pinter, London) 34.
2. Don Lamberton, 1983, Information economics and technological change, In: Don Lamberton, Stuart Macdonald and Tom Mandeville, eds, *The Trouble with Technology: Exploration in the Process of Technological Change* (London: Frances Pinter), 83.
3. Rich Karigaard, 1995, Explaining growth, *Forbes*, (5 June): 9.
4. Robert M. Solow, 1956, A contribution to the theory of economic growth, *Quarterly Journal of Economics*, (February):65-94.
5. Robert M. Solow, 1957, Technical change and the aggregate production function, *Review of Economics and Statistics*, 39:312-320.
6. Paul Romer, 1987, Crazy explanations for the productivity slowdown, *NBER Macroeconomics Annual* 2:163-201.
7. Sigmund Freud, 1930, *Civilisation and Its Discontents* (London, W.W. Norton) and Alvin Toffler, 1970, *Future Shock* (Random House, New York).
8. Meheroo Jussawalla, 1983, Trade, technology transfer, and development. In: Don Lamberton, Stuart Macdonald and Tom Mandeville, eds, *The Trouble with Technology: Exploration in the Process of Technological Change* (Frances Pinter, London) 134-154.
9. Ian Reinecke, 1983, The trouble with techno speak. In: Don Lamberton, Stuart Macdonald, and Tom Mandeville, eds, *The Trouble with Technology: Exploration in the Process of Technological Change* (Frances Pinter, London) 194.
10. Ian Reinecke, *ibid.*

CHAPTER 5

Comment: David Cracknell

The argument here could be strengthened by presenting some reasons why government statisticians might want to measure technological change. In the UK price inflation measures were developed to track the cost of living and (presumably) to moderate wage demands. It might be argued that some measure of product innovation effectiveness could have the same effect on capital demands, and might influence capital/labour substitution.

It is possible to model demand for new services through quality adjusted prices if you get the quality adjustment right. In the example of micro-electronics, you would have to do this by reference to the final consumer products. It should be possible to predict the impact of electronics on the cost, quality and features of watches for instance to assess the impact on final demand.

Is it perceived rather than actual user value that is the important factor in take-up of new services. A customer buying a new watch with additional features is, for instance, unlikely to be able to make a clear judgment about the use he will be able to make of them in practice, but will find some subjective way of placing a value upon them.

CHAPTER 6

Market structures in telecommunications: the agenda

John Nightingale

1. Market or industry structure

The conventional definition of market structure

The standard conception of market structure was developed mainly by the American industrial organization economists, Edward Mason, J.M Clark, and Joe Bain. Their structure-conduct-performance framework has provided a durable means of analysis, but the results have never been totally satisfying. As F.M. Scherer pointed out long ago, the direction of causality implied is by no means the only one (Scherer, 1970:5-6), the econometrics have done little more than not contradict the impressions of less than casual observers (Scherer and Ross, 1990:ch. 4), and the generalizations are few and not very strong (Sutton, 1991:ch. 14). The so-called New Industrial Economics has left generalization even further behind (Brander, 1992). Game theory has been fruitful in *ex post* rationalization, but fruitless in *ex ante* prediction, and incapable of any but the most obvious generalizations, ones made decades earlier by the obsolete "old" industrial economics.

The two generalizations for which Sutton found support were that higher profit rates were associated with higher concentration, and that high sunk costs were associated with high concentration. To these we can add Scherer's generalization that the rate of technological progress is associated with concentration not too low and not too high (Scherer and Ross, 1990:660).

While this might seem meagre pickings, they are of immediate relevance to the telecommunications industries, all of which seem to be characterized by high sunk costs and high rates of technological progress. But if all the conventional framework can deliver is a verdict of "not surprising" it will fail to enlighten either economists or policy-makers. The research agenda must move on beyond these generalizations if it is to contribute insights. The beginning made by the standard framework can be built on by extending that system to deal with what seems to be central to observers of telecommunications, i.e., the dynamics of large corporations interacting under conditions of partially self imposed rapid technological transformations where network externalities dominate the inter-relationships between these corporations. In so doing, the models lose their neoclassical flavour and find themselves in the realms of evolutionary models in which the arrow of time is of the essence, paths of growth and change are the phenomena they explain, history is important, the processes being observed usually show path dependence and small shocks to these economic systems can lead to changes in rates and directions of change that would not have been observed without such shocks.

Static models of market structure

John Sutton (1991) provides perhaps the simplest static framework within which the industries of the telecommunications sector may be analysed. His two stage game takes into account the decision to enter as the long run structural determinant, while the decision to use one or another strategy yields short term competitive behaviour.

If the entry[1] decision is the sole element of sunk costs, then his exogenous sunk costs model is appropriate. Under this model, the long run equilibrium structure of the industry is determined by the relationship between sunk costs and market size. The greater are sunk costs imposed on the players by exogenously determined technology, the greater the market size required to allow market concentration to be diluted sufficiently to allow the second stage game to result in socially acceptable price cost margins. Whichever second stage strategy[2] is observed, depends partly on first stage structural features. Within this research programme, the task of the observing economist is to define the sunk cost condition of the industry. This is conceptually relatively simple. The results it promises are considerable. It may indeed be a very substantial beginning to understanding telecommunications industries. But is it likely to be enough?

The drawbacks begin with, first, the long run equilibrium nature of the model: It is based on the assumption that our observations are of the final state of rest which is the outcome of the game. But we are denied observation of such a long run. Indeed, much of the interest in the sector is that it appears to be far from equilibrium, and its path of change rather than its equilibrium characteristics are of direct and immediate scientific and policy interest. Waiting for equilibrium to emerge in order to observe it may be like waiting for Godot.

The second drawback is that sunk costs do not appear to be wholly exogenous. Advertising and marketing, quality of service on both the engineers' technical definition and on many additional ones, are important endogenous sunk costs as are the level of R&D spending, not to mention its direction depending on the firm's strategic goals. In Sutton's second model the sunk costs are endogenous, so that the strategic game of the incumbent players is one of committing funds to a chosen quality attribute level, not merely the entry decision. This in turn commits them to market behaviours. The sunk cost here is in a vertical quality attribute, so that demand shifts upward with investment spending. But spending is independent of volume of sales, and is again an equilibrium value chosen prior to playing the second stage. The analogue in telecommunications may be the decision to invest in one or another quality of service level in a network. This is a very limited conception of endogeneity of sunk costs. It excludes both the question of direction of spending — advertising, R&D, infrastructure quality, which itself has obvious strategic importance, as well as actively manipulating commitments within the second stage game of price and output strategy. Sutton's model cannot allow this, as the two games are not essentially time bound, but occur simultaneously as decisions to create the system.

However, even in the limited regime of his model, he finds an extremely robust result which is surprisingly nihilistic. It is that "(i)f it is possible to enhance customers' willingness-to-pay for a given product to some minimal degree by way of a proportionate increase in fixed cost (with either no increase or only a small increase in unit variable costs), then the industry will not converge to a fragmented structure, however large the market becomes" (Sutton, *ibid.*:47). This result continually leaps out, to be suppressed by some or other special and particular condition which will prevent a persistent advantage to size emerging in the equilibrium condition (*ibid.*:46-48). The implication is that allowing this aspect of reality into the model destroys any chance of associating static exogenous features with performance. It doesn't matter that the market is extremely large, or that markets are open to entry, or that more than two players are incumbent, the equilibrium properties remain stubbornly and indeterminately concentrated, price and quality outcomes sensitively dependent on precise and generally unknown, even unknowable (commercially secret) aspects of firms' choices and the interrelationships between firms. There is more than a hint that in such industrial structures, so common in the modern industrial economy, equilibrium is not an observation but at best an attractor as in a chaos model. At worst it is a set of differing attractors around which and between which actual market structures oscillate, firm's fortunes waxing and waning as market success is enjoyed, then lost.

The implication for the use of conventional market structure models of the kind labelled "old" industrial economics, is that they are too simple. For the "new" industrial economics the implication is that analysis is doomed to special cases, the search for the Holy Grail of a general explanatory and predictive theory is forever beyond their grasp.

Recent controversies in the theory of the firm have seen the "New Institutionalist" contracts view of the firm and the industry (Coase/Williamson transactions costs economics, together with Agency Theory) pitted against the competency or capabilities view of the firm taken by evolutionary economists (Marshall, 1890; Nelson and Winter, 1982; Eliasson, 1991) and resource based theories of strategic management (Langlois, 1995). The controversy may well be on the way to resolution with recent developments in the theory. Lazonick, Langlois, and Loasby, in particular, have contributed to this, as shown below. This alliterative trio have divergent views but the synthesis suggested by Loasby (1991) gives a hint as to a possible resolution. This resolution has important implications for the agenda of research into the structures which we see emerging in the telecommunications sector. While perhaps more relevant to the topic of networking in this volume it cannot be ignored in the context of market structures, as it provides a link between the traditional concerns of orthodox neoclassical economics and those of evolutionary, time bound, economics which I hope to show to be of great promise in terms of explaining and predicting directions of future change in market structures in this sector.

The "new institutional economics": agency theory and transactions costs economics

These extensions of orthodoxy are fundamentally attempts to explain the existence of the firm, given a commitment to methodological individualism (Winter, 1991). They turn attention away from the firm as a productive enterprise in order to focus on the question of exchange: how may exchanges be organized by agents anxious to benefit from their individual capabilities to engage in productive services? Fully independent individuals are seen to fail to enjoy benefits due to opportunism combined with at least asymmetric knowledge endowments. In the transactions cost version complexity prevents agents knowing enough to enter into complete contracts, these being necessary due to the sinking of investment funds into specific assets.

This picture of the capitalist firm as arising to resolve merely exchange problems has been subjected to damaging critique by a range of scholars, including Charles Perrow (1986), William Lazonick (1991), Sidney Winter (1991), Neil Kay (1992), Richard Langlois (1995), and Brian Loasby (1994). However, there are aspects of these ideas deserving of preservation, as Loasby points out (*ibid.*). Agency problems are part of organizational behaviour; transactions costs are a significant if not dominant influence on institutional arrangements. For our present purposes these theories provide an important element in the research agenda in that telecommunications services are central to exchange media in the global economy. These services embody exchange as an institution. Analysis of the institutions of telecommunications services in terms of their facilitation of exchange is a task for agency and transactions costs economics, or at least for analysis informed by the best insights of these theories.

The use of these ideas comes in at least three guises. The first is in understanding the alternative forms of vertical relation observed in the telecommunications services industries: local providers interconnect with regional, national, international. How are these exchanges organized? Is the contracting question central or peripheral in illuminating the forms being observed? Has the rise and decline of different forms of contracting these exchanges been due to agency or transactional problems?

The second aspect is understanding the demand for telecommunications services as demand for contracting media, these demands in turn causing the vertical relations of the telecommunications services industries. This aspect of the matter may be better dealt with under the demand chapter, but it should be noted that the nature of the vertical relations is clearly a response to demand for services. It is also the case that integration between service provision and service users is common as value added provision by telecommunications service providers. To what extent can this integration be explained simply by the contracting problems between telecommunications providers and value added service providers such as cable television, interactive information and transactions services?

The third aspect is that of regulation as a means of enforcing allocative and productive efficiency where markets are seen to fail (Laffont and Tirole, 1993). It attempts to draw out the implications of individual optimisation for the productive efficiency of organizations, whether the focus is on agency problems within an organization or on agency problems within the system bounded by the regulator and the firm under regulation. In this latter context agency theory has an important role in the analysis of emergent market structures. The worry is that thus far, and despite an enormous research effort, the results have been disappointing. Generalizations are few, limited to the expectation that productive as well as allocative efficiency will be compromised by agency problems, and the desirability of some form of risk sharing between principal (or regulator) and agent (or firm) as a means of reducing the agency problem of information asymmetry. Specific models have specific implications, but all too often these are the outcome of auxiliary hypotheses which do not bear application. The careful building of directly applied models can be of use in suggesting directions of institutional reform in which organizations and regulated industries may improve productive and allocative efficiency. But that care is difficult to exercise, as the precise structural effects of auxiliary assumptions can be quite opaque. When it is further noted that these models assume substantive rationality on the part of the agent, with the principal attempting to infer what that true knowledge is, the field of application is again circumscribed. The choice of assumptions is critical to the model's outcomes, sometimes sensitively so.

Evolutionary or competency theories of the firm

On its own, the agency and transactions costs approach cannot be a complete explanation of the institutions observed as it ignores competencies in production of services. Competency, or evolutionary theories of the firm, are closely related to resource based strategic management theory, deriving largely from the work of H. Igor Ansoff (e.g., his 1982), but for economists that of Penrose (1959) and Marris (1964) is central (Langlois, 1995). These can be applied at the level of the firm, the network (Antonelli, 1994, and this volume), the sector, mesoeconomy or filière (Antonelli, Petit and Tahar, 1992), and the global economy (Silverberg and Verspagen, 1994). Fascinating to economists is the clear intellectual thread which this analysis has from Marshall's *Principles* (Lazonick, 1991, for a particularly clear demonstration, but see also Loasby, 1991).

As Lazonick (1991) points out, the firm is a core of competence which enjoys productive economies internal to itself as a coordinating unit. Differences between degrees of vertical integration are centrally differences in internalization of these economies. Indeed, much vertical integration is at the cost of transactional efficiency. The contracting problems of the complex firm are worth bearing due to the productive economies enjoyed by that more complex firm. But as Langlois (1995) and Loasby (1994) point out, the market is also an institution which embodies coordination of production capabilities. Lazonick's own examples of the nature of the 19th century labour market in the industrial districts of England illustrates this point well. The labour markets of each district not only gave form to contracts between master and craftsman, but also organized knowledge about the craft (Lazonick, *ibid.*, ch. 1; Loasby, 1994). So it is in telecommunications services, that choices of internalization and the nature of the internalization or integration are informed by the need to coordinate productive resources into the systems that effect production. This is another theme also central to the network concept covered in another chapter of this volume.

The great advantage of evolutionary theory is in its focus on diversity and experimentation, on the nature of economic change and the process of growth and structural transformation. The direction in which such theory points is explanation of historical processes of change. Choices made by entities at different points in time are followed through to consequences for market success or failure. The making of choices in an uncertain environment, one in which ignorance is endemic, cannot be any but a process of experimentation. As an example of this form of modelling, Eliasson (1991) is explicitly concerned with the consequences of experimentation for market structure. The consequences of attempts at directing or controlling experimentation, by regulators, or by the restrictive premises of the firm, is another significant area of research which impinges on the regulation of telecommunications during such a time of turbulence.

Distributional consequences of these changes are another key issue for the research agenda. The contribution made by evolutionary theories here is one of critique of the static analysis of distributional effects. Yet again the difficulty of distinguishing transitional and turbulent aspects of these changes from long term trajectories of change come to the fore. Letting markets rip into experiment and attempts at exploitation of transitional rents to innovation is inevitably going to lead to non-Pareto improvements. Domestic household users of telecommunications seem to be least prominent prospects for great gains in such a context. Universal service would be the first concept to be abandoned by firms freed from regulatory constraints. Analysis of the levels at which universal service obligations might be applied during transitional periods is clearly required. But it is not clear that the orthodox analysis would be up to the job, as the static equilibrium framework appropriate to a steady state may give quite poor guidance during the turbulence. The catch cry of "cross subsidy" central to that analysis fails to account for the effects external to the cross subsidized service, effects which are captured by a broader analysis which internalizes those effects. For example, is the loss made by public pay phones greater than the gains to the pay phone provider from having its network customers access pay phone users? An example of an obvious beneficiary of pay phone access are taxi services, but there are many others. Should these beneficiaries subsidize pay phones? Or should pay phones be priced to pay their way? Does a price exist at every point in time as the turbulence of experimentation proceeds which would allow pay phones to operate profitably? Answers to these questions are not obvious, probably not to firms, let alone regulators or researchers.

But do these distributional issues have a direct link with market structures? On the one hand it can be argued by the social engineering branch of regulatory policy-makers that tax and subsidy solutions are available to enable social policy to cope with the question of universal service obligations under any structure that happens to evolve, or is engineered to evolve in pursuit of other policy goals. This is, broadly speaking, the solution adopted by policy-makers in a number of jurisdictions in the United States, and in Australia (Schankerman, 1996). While the dominant firm supplies the services under social obligation, other firms contribute to financing that supply roughly in proportion to their market share. On the other hand, social engineering is never more than rough and ready, fraught with unintended consequences and subject to political pressures from many directions. It may be unreasonable to expect a market structured for allocative and productive efficiency, or for rapid economic growth, to deliver policy-makers' goals. In particular, the very phenomenon of rapid change may carry with it consequences of resistance, distrust and political disruption, as the apparent losers fight to defend their past benefits. Even if change is actually Pareto improving this might not be obvious or inevitable, fanning the flames of resistance. Price rebalancing is one element of telecommunications reforms that has been delayed by these concerns. A slower pace of change, and an adequate consideration of distributional issues such that minimal regulatory interference on distributional grounds is necessary, may well create a more politically sustainable market structure in a democratic environment. This issue is one requiring attention on the research agenda, perhaps not mainly from economists.

Richardson's notion of multi-variate institutions which carry out the coordination of production and exchange

Loasby (1994) takes us back to G.B. Richardson's 1972 paper (in Richardson, 1990) which anticipates virtually the entire controversy between Williamson and the competence theorists by suggesting that the institutions of production and exchange have to be considered together as means chosen by decentralized capitalist decision-makers in their pursuit of profit, dominance, and whatever else they desire.

The various institutions we observe are solutions to these problems of management or coordination. The market, seen as impersonal, arms length, is an extreme on the dimension of openness to contestants. This extreme is the orthodox conception of a market, in which suppliers stand ready to sell to whomsoever, and would be buyers make their offers to this impersonal market. Suffice it to say that such markets are not of great importance in the telecommunications sector. The markets whose structure is of interest are those for which relationships which persist over time are necessary to the efficient coordination of supplies to users, and of the production of those supplies by suppliers. Again, we enter the world of networks of relationships defined by the temporal interconnectedness of production of the stream of services supplied by the sector to its end users.

I hope that I have shown that the work of Loasby, Lazonick and Langlois suggest that both exchange and production questions can be approached consistently in a framework of historical and institutional analysis of economic forces within times of great change and turbulence. It is this which makes their work of relevance to the study of telecommunications in the present era.

Extensions of the conventional in dynamic systems

Thus far discussion has skirted around the question of time and path dependence, and the place of equilibrium analysis in a world of endogenously changing demand and technology. The Sutton model shows the inadequacy of the static approach to an industry/market system undergoing rapid change as is telecommunications during the present decades. We neither know the characteristics of an equilibrium system should it emerge at some time distant from our own, nor would we care too much about that equilibrium, as time horizons of policy-makers are not located far enough away in the future. Our interest is in the properties of the system as we see it today, and as we imagine it developing within a policy time frame.

An explicitly historical and evolutionary perspective is required to provide a rigorous framework within which to conceptualize the present rapid changes in market structure, many of which appear to be random responses to uncertainty. The concept "market turbulence" is perhaps an initial marketer's response to what seems to be randomness. Markets in which competitive advantage shifts markedly with changes in offerings, with the stream of competitive moves and adaptations to emerging market conditions, are far from equilibrium, and require analysis which can model that process or set of processes.

Models capable of dealing with such turbulence must first recognize the experimental nature of firms' competitive moves in the market place. Information does not exist on the formation of demand, because in too many cases of service offerings the potential users are in the same sea of uncertainty about the offerings as the providers. Neither users nor providers know, nor can they know, what it is that a particular service on offer may be appropriated for by experimental users. The Minitel is an excellent example of such a service. The Internet is another example of a set of services provided in this case by a decentralized set of user/providers who had no idea what they were doing, and to a considerable extent still do not know. Modelling experimentation is a strength of Eliasson's work (e.g., 1991), as it is of Nelson and Winter (1982), Silverberg (e.g., 1994), Dosi (e.g., 1988) and the other workers in the technical change field outside of the neoclassical world of known production functions. Faced by an adverse environmental change or by uncertainty requiring some response to limit its adverse effects on management's ability to run its firm, there is search in directions acceptable to the firm (Perrow, 1986, ch. 4 has a good discussion of the "premises of the organization", which direct search amongst other things).

But experimentation and adaptation is necessarily only a part of an organization's life. There must be some coherence to the firm if it is to be capable of directing and coordinating activity. It is this coherence which defines the firm, or the firm-like set of institutions which constitute the production unit. The unchanging characteristics of this unit constitutes the feature on which selection pressures can bear to reward or punish the unit in the competitive system. Competition is both in the marketplace for the services it offers users, and in the regulatory and political marketplace for favours and grants of monopoly or right of entry. The identification of continuing selection characteristics in coherent productive entities should be a major element in this research programme. And while much work has been done by organization and management science to identify means of responding to pressure, the same may not be as true of efforts to identify what continuing features may be identified as favourable selection characteristics. Thus far, evolutionary economists' models tend to be built on simplistic assumptions that input ratios define a technique, and input/output ratios define the level of productivity. They model innovation as draws from an urn of possible alternative techniques, with various probabilities of improvement in productivity, and of how different the input ratios might be for the new technique. Thus the selection variable is the individual firm profit level given the input cost environment of all competing firms, where diversity derives from different techniques and productivities. Clearly much effort has to go into modelling selection characteristics.

The behaviour of systems comprising competing and cooperating groups is not well understood, but is the subject of an enormously exciting research effort. The simulation studies of Nelson and Winter gave this programme its first major impact on the economics profession. But there are many intellectual roots to the evolutionary project in economics. Geoffrey Hodgson's review of the history of evolutionary thought in economics (1993) shows the breadth of the programme. The levels on which this programme must be developed are many, including the macroeconomic level of Nelson and Winter (1982), Eliasson (1991), and Silverberg and Verspagen (1994). The level at which the evolution of market structures, technologies, and indeed regulatory regimes, in the telecommunications sector may be studied is the level of the filière. Antonelli, Petit and Tahar (1992) is an example of such this form of analysis, in their case the cotton filière.

In such analysis the central concern is not equilibrium nor comparative statics, but explanation of the historical path of growth and change. William Lazonick would have this subject to the criteria of rigorous historical analysis (1991:304). But it must go further, first in demonstrating the causal chains and whether those chains were accidents of history or were, for example, the inevitable consequence of the structure of the system, as in "edge of chaos" models (Lane, 1993). Secondly, analysis must attempt to specify alternative futures and the contingencies of those alternatives. This is as close as we may come to "comparative statics", for the contingent events are in general irreversible in their effects: a tax increase will affect the choices made by businesses, but reversal of the increase will not take those businesses back to where they began. They have learned new tactics, new methods, in coping with the increase, and will not unlearn them on removal of the tax.

Industry structure and market structure as an element in institutional structure that characterizes provision of telecommunications services

The exogenous features of industries in the telecommunications sector include not only those highlighted by Sutton's models, that is, the nature and extent of sunk costs. They also include the broad trajectories of core technologies whose current rapid evolution can be understood only very partially as endogenous to the telecommunications sector. The impact of these changes has to be explained. Why have such changes been taken up, what effect have they had, are they having, on industry structure?

The disappearance of monopoly, the belief that natural monopoly is no longer the case for delivery of telecommunications services, can be said to flow from the changes of technology in microelectronics, computing and other complementary technologies. But does this belief feed upon itself? Does the exogenous trajectory, or the perception of that trajectory, lead to endogenous change, to direction of research along the trajectory which appears to hold greatest opportunity? And does public policy directed toward creating contested markets rather than monopoly markets further drive research effort designed to allow entry to be sustainable?

The institutions of the telecommunications sector are clearly formed by both what is seen as possible, and what is seen as desirable. The visionaries whose perceptions count are both within the industry, policy-making bodies and the academy. Indeed, the value placed on creating contested, or competitive, markets, is attributed by some (especially management within the once natural monopolies) to economic theory. The possibility of contested markets, given the technological environment, is then available to fulfil the dream of that desirability. But alternatives are available. Monopoly provision of much social infrastructure is still common in advanced western economies. There are still those who would argue that efficient provision of a "National Information Network" would be by central control, in much the same manner as provision of a monetary system by monetary institutions of national government. It is thus hard to argue for a deterministic view of industry structure as the inevitable outcome of exogenous technological factors. The task of explaining the structures, and their trajectories of change is thus one full of interest and potential for surprise.

2. The *status quo ante*

Prior to the explosion of technological and institutional change the PTTs and their suppliers were linked in ways easily understood by the standard framework of industrial organization economics. The distinction between service suppliers and equipment which they used, and indeed supplied to their subscribers was important but straightforward. Different degrees of vertical integration of various forms were observed, from completely unintegrated to completely integrated. However, the more recent distinction between facilities and services was of virtually no significance. The telephone system was for voices only. Telegrams and telex were a distinct and different service provided also by the PTT, except in a few cases such as the USA. But even there, the service and the facility on which it was provided was the same. The nature of the sector was very much simpler in the days where it approximated a steady state, in which the major change element was increasing market penetration as the telephone diffused into the final state of its market niche.

Monopoly structures in service provision

Littlechild (1979) is an economics textbook for telephone engineers. It presupposes the regulated monopoly structure of the post-war world. The analysis is of service provision over a network infrastructure which is coterminous with the service provision.

There is little of contemporary interest in the issues discussed, in the sense that the analysis is unsurprising and the discussion of regulation is on the assumption that it is in the public interest. Whether the regulator is independent of the service provider or not is irrelevant. The public service department PTT will behave just as the American FCC would attempt to induce the Bell system to behave. The analysis of demand is as though demand were predominantly from households rather than from the business sector. Technology is assumed to determine natural monopoly as efficient, and the only appropriate institutional arrangement for the provision of telephone services.

Clearly the contemporary analysis of natural monopoly is rather different from that of twenty years ago. Laffont and Tirole (1993) represents the current "state of the art" in the analysis of regulated natural monopoly, much of which was developed in the telecommunications context (for example, Sharkey (1982) is a product of Bell Labs). While there is much of relevance, monopoly as a phenomenon is now receding in importance as technological change pushes natural monopoly back to fewer and fewer markets, and as the importance of networks of markets/industries grows. The research agenda has clearly moved on.

Bilateral oligopoly in equipment provision

Provision of equipment for the PTTs was once a matter of supply of all the physical material facilities required for the network, from telephones in subscriber premises to ceramic insulators on telephone poles. Thus the PTTs contracted with their suppliers in a market that was a classic example of bilateral oligopoly. It is hardly surprising that a considerable degree of vertical integration was observed. And where vertical integration did not prevail, relations between buyer and seller were typically long established ones, with much effort devoted to ensuring continuity of supply by the PTTs and certainty of continuing sales by the suppliers.

There were at least two motives for some variety of vertical relationship stronger than spot market deals. The first is ensuring continued supply of specialized outputs, specific to the particular PTT. In the case of Australia, the immediate post-war period saw an effort at import replacement to ensure minimum reliance on offshore supplies in a time of considerable uncertainty about those supplies, especially in case of superpower conflict. The second is the panoply of reasons for vertical integration in bilateral oligopoly, that is, reduction of contracting costs, avoidance of disputes about division of surplus, protection of intellectual property, and so on (Buckley and Casson, 1991, is a classic exposition in the multinational context, again important for telecommunications).

The explanation of vertical integration, the theory which suggests vertical relations will be strong, if not wholly integrated, is compelling, so much so that the problem for the research agenda is in explaining why it is that competitive tendering exists at all in the supply of physical and intellectual equipment to the successors to PTTs. In the earlier period under discussion here, integration generally meant protection for local industry, generally a single major supplier of goods such as cable, exchanges, handsets. In Australia, a two supplier principle was applied to further ensure competitive supply without full integration. Thus major international equipment firms were induced to invest in production in Australia on promise that the Postmaster General's Department would purchase predominantly from local suppliers. At one stage, the proportion of purchases for the telephone system which was local value added was above 80% (Industries Assistance Commission, 1984). But the PTT owned no significant suppliers. This form of integration was a national policy applied directly by the PTT, with the collusion of government, a policy entirely congruent with the post-war demand for security of supplies. A thorough investigation of the history of this period of Australia's PTT's development is yet to be done, but it would be surprising if there was an open market within Australia for any company which decided to set up production facilities for specific telephone system components.

Again it must be emphasized, the industries buying and selling telecommunications equipment and services were clearly defined in two parts: PTTs supplying telecommunications services monopolistically and equipment suppliers serving those PTTs. The notion of facilities separate from services carried was not defined until the middle 1980s where it emerged initially as a distinction between an integrated facility/basic service (POTS) industry, and suppliers of so-called value added services. And there is where the really messy difficulties began. The engineering analysis of facilities (an analysis which doubtless predated the economic debate by decades) into a number of levels, from the physical cables and wave guides, signal protocols for the carrier stream, up to the level of the protocols that handsets required to access analogue or digital information streams, had opened the floodgates to the present complexities which make physical infrastructure provision separable from carrier wave provision, from network provision, from services provided on the networks. There are, of course, those who still wonder at the importance of the distinction between facilities and services. But the greatest wonderment must surely be as to where lines are appropriately drawn? Can facilities mean simply the cables and switches giving access to exchanges? Or does it include the public exchanges as well, given that there is such a thing as a public switched network? Or does facilities include also everything that allows a customer to connect to any network? With or without POTS? How far can vertical disintegration be sensibly argued? The same arguments are taking place in other network provisions, such as railways. The most radical experiments in vertical disintegration in railways, e.g. in the UK where the permanent way is controlled by one institution, and different classes of carrier by others, are in their early days. There is at least the possibility that such disintegration has given away too much coordinating power, but evaluation must await some many years of the experiment. Voices are heard which doubt the possibility of maintaining any particular set of distinctions between facilities and services. These voices tell of firms initially merely service providers which soon enough invest in their own facilities of one or another form. The eagerness of Pay TV providers to control their own cables, rather than merely renting space on other firms' cables is another straw in the wind.

Regulatory certainty and national policy prescribed structure and performance

With the possible exception of the USA, the old world of POTS was one of national policy driving the monopolistic service provider. It was a public service, in the interests of economic and strategic national goals. Subventions from the national budget were made, offset by user charges based on criteria familiar from a previous century of postal service history. Even in North America belief in the naturalness of monopoly was accepted, even if this belief had been carefully manufactured by Theordore Vail as a means of ensuring the dominance of the Bell System (Bornholz and Evans, 1983).

National policy also determined the extent of integration of the equipment supply industry, as pointed out above. It was thus the case that market structure was based on beliefs about what was appropriate provision of a national service, "obviously" a natural monopoly (if predating the economists' definition and analysis), generally under national control as an important part of the nation's security, not usually seen as a part of a market economy itself but the servant of that economy. Economic and market analysis was not important to its provision, but sound engineering practice was. Technology driven criteria of performance were at the centre of national goals, as Littlechild's analysis of the economics of such a system showed (Littlechild, 1979). That economic analysis was a minor part of the criteria for service provision, the major part being the national goals set by governments for network diffusion.

Rigorous historical analysis, as Lazonick demands (1991), may not yet be complete. But that history will doubtless reveal more complexity than is now apparent in what appears from present perspectives to be a remarkably simple environment for the PTT. The research agenda must include such an analysis if the present is to be well understood. Snow (1995) indicates that some of this has begun. Moyal's work on Australia's telephone system (1984) is a narrative which must be extended into the public policy area as historical analysis of broad themes of Australian economic and social history is elaborated (Rimmer, 1988; Livingston, 1993 are examples).

3. The forces for change in the second half of this century

Anti-trust and competition policy in the US and later other countries

America's AT&T has lived most of its life under the jurisdiction of one or another part of anti-trust policy, as well as of FCC regulation, not to mention state public utility regulators in the various jurisdictions of the nation. It is clear that AT&T has not only been regulated in all these various ways but that it has built strategy on the basis of regulation, and probably sought policy directions which would suit its own purposes. To what extent was monopoly natural, to what extent was it a creation of AT&T, of Theodore Vail, to justify its monopoly position?

As Bornholz and Evans (1983) tell the story it is one of the Bell company intent on controlling the market, from its patent protected era prior to 1893-94 until Judge Greene's modified final judgement of 1982. The structure of the system, evolving from the licensing agency which originally held the patents and its licensees, with freedom of entry post patent, meant a plethora of operating companies which AT&T struggled to control by all means at its disposal, mainly financial and legal (*ibid.*:8-14). The goal of AT&T of monopolizing telephones was not achieved until 1934, despite all the stratagems at the disposal of a powerful corporation backed by such financiers as J.P. Morgan. Bornholz and Evans (*ibid.*) argue that AT&T achieved its modern form in the 1907 financial restructuring which saw Morgan's interests and Vail in control, and the previous regime of Boston capitalists and the management from the patent protected era replaced. They claim that Vail's policies dating from that period can be seen in unchanged form through to the 1980s when the divestiture took place. The inference they draw from this five decade long struggle to create and maintain the monopoly is that it was a most unnatural monopoly. But this inference is not justified. Let me argue that natural monopoly may have obtained between the 1880s and the 1970s or so. This argument will show that much more historical research remains before any firm conclusions can be drawn.

The existence of competitors, especially rival firms tying up local markets not served by the Bell system, but also rivals in large markets, and rival long lines, merely means that there is a fight for supremacy. The telephone was a rapidly diffusing service, the industry highly immature. As Bornholz and Evans point out (*ibid.*:12), as late as the turn of the century Vail, general manager from 1879 to 1887, stockholder then President under the Morgan regime, "recognized that 'the worst of the opposition has come from a lack of facilities afforded by our companies, that is either no service, or poor service'" (*ibid.*). In other words, the question was not "what is the efficient means of provision" but how can any provision be extended in a rapidly growing market? The problem of financing rapid expansion, rather than that of achieving minimum cost of service provision, is illustrated by the historical process of the Bell system's efforts to create a powerful and unchallenged place in American telecommunications. Even at a relatively late stage in Bell's absorption of the independents, the fact that the process was continuing is not *prima facie* evidence that the institutional arrangements of the industry were becoming more or less efficient. The change of policy from aggressive expansion and refusal to deal in any way with independents, prior to the change of control, to absorption and co-option after, may merely indicate that the Morgan interests judged that the least costly way of achieving the monopoly was the new policy. The old meant many wars of attrition and heavy investment in fixed assets which might be redundant after the rival independent was forced to accept defeat. The new meant investment in much more fungible assets. The venerable Chicago argument that predatory pricing is irrational may be relevant here. The old policy implied reliance on the rationality of investors in the independents to see that they could not compete with Bell, a reliance that had to be tested over a long period of oligopolistic rivalry to convince them that profits were not available in the face of the superior efficiency of the Bell system. If investors were not convinced, if they thought they could compete, that they could operate more cheaply than Bell, the fight would go on. Indeed there is no need for it to ever end, with both making acceptable profits, but at average costs higher than either could achieve in the absence of the other from the market. There is no market imperative to see one or the other off in this situation, and the sub-optimal duopoly provision can continue indefinitely. The new policy allowed negotiation and deal making which results in both entities, both sets of investors, earning far improved profits. Whether consumers of telecommunications services enjoyed better or cheaper service is another question entirely, a question the anti-trust authorities did consider in their oversight of Bell, but not one which allowed of definitive answers.

The analysis of these questions in Bornholz and Evans goes some way, but not far enough, in answering them. Their theoretical perspective is that of equilibrium theory, so they have in mind that at each point in time the policies being followed, and the *status quo* of the day, were characterizations of equilibrium. If, on the other hand, we characterize policies and *status quo* as historical artefacts of disequilibrium the implications they draw are seen to be too strong. Their analysis (*ibid.*:14-34) is couched in terms of strategy judged in light of the various firms' costs being in fact the minimum average costs for the relevant outputs. They go so far as to draw on the orthodox analysis of natural monopoly to imply that Vail's arguments that one system can serve the community more efficiently than two, is necessarily wrong. But I would argue we can draw no such conclusion. Vail may well have been wrong, but not because the evidence adduced shows telephone service to be not a natural monopoly. That evidence merely shows that two telephone systems serving the one community were financially viable at that time, given pricing policies chosen by Bell in response to competition, given the geographical distribution of the competing systems (often in competition directly only in the CBD), given policies of the two systems regarding further extension of service and diffusion into the residential and rural markets, given re-investment policies for internally generated funds, and so on. Argument based on relative pricing levels in contested as against uncontested markets would seem to be not very relevant to the question. Of course the monopolist will attempt to milk the captive market to support dividend, investment and market contesting policies. Of course it may be better for consumers to reside in a contested market area, and of course it may be better for development of services to have contested markets, *even at some cost to productive and allocative efficiency.*

Especially historically presumptuous are statements such as that at (*ibid.*:31):

> If local service had been a natural monopoly, the existing Bell exchanges could have captured the entire local market by expanding their systems. The expansion would have raised their costs, on the one hand, and increased the value of their service, on the other hand ... (T)hey would have been able to set prices that would have enticed subscribers to join their systems rather than the independent systems. The fact that, even after the Bell exchanges slashed their prices, by about half, to the levels offered by the independents, they were unable to establish a monopoly suggests that the addition of subscribers increased the cost of service more rapidly than the value of service. Groups of subscribers obviously found it to be in their self-interest not to join other groups of subscribers connected with the Bell System.

The flaws in this argument include the following: 1. Subscribers respond not merely to price, but to other aspects of the offer of service, including knowledge that monopoly markets bore a much higher toll, so best to keep it contested; 2. Bell policy post-1907 was not one of slash and burn, but one of financial attrition, and co-option. They did not wish to capture the whole market at great expense, as I pointed out above from Bornholz and Evans (*ibid.*:13); 3. The immediate consequences of capturing an independent may well be an increase of average cost of service, which would make the capture tactic more costly again, but there must be evidence that this is a long run level of cost before it can be claimed that the cost of service by a monopolist is no lower than of two contested systems. 4. Evidence that returns to scale were less than unity is relevant to the internal arrangements of firms rather than to the organization of the system as a whole. The evidence relates to the costs per line of an exchange, which is one element of the firm's costs. It could be taken more seriously if the size of exchanges was seen to be chosen at particular limits. The period during which this limitation applied is not clear, but it may be that the period did not last much beyond the manual exchange era.

Much more, and more rigorous, historical analysis will be required before the acceptance of argument that American telecommunications did or did not exhibit the characteristics of natural monopoly at any particular stage of its development.

The argument has often been made that AT&T made its choices of its permitted activities, moving from prohibition from involvement in the computer industry by relinquishing its local and regional operating subsidiaries in the landmark "modified final judgement" of Judge Greene in 1982 (Evans and Rothschild, 1983:157). But in this breaking up of the Bell System came the first evidence of the great structural changes that were to transform the sector world wide. To what extent Judge Greene's decision was purely an outcome of processes familiar to anti-trust enforcement as against any new direction of policy is an argument which will remain on the agenda for some little time. However, the break up of Bell signalled the era of contestable oligopoly had officially been recognized. Competition in long lines, already of marginal importance, was now institutionally favoured with a complex set of regulations designed to foster entry.

It is true that such competition could not be fostered if the technologies for competition did not exist. But the existence of alternative and cost effective infrastructure was but a shaky possibility for entrants until favourable regulatory treatment allowed the development of those technologies via investment by those potential entrants. It was the policies of the American anti-trust authorities which provided those conditions, necessary but not sufficient for competitors of the dominant firm to grow from vulnerable entrants to viable independent competitors.

In countries other than the USA political and ideological forces were also a crucial element in breaking the old monopolies. The UK had Mrs Thatcher's drive for privatization and competition in government business enterprise. The privatization of British Telecom, first and greatest of the privatizations, was accompanied by the entry of Mercury, and later of other carriers from the cable television and related industries. The motive for these reforms was not simply economic efficiency but included ideological enthusiasm for private ownership and the rhetoric of competition. A fundamental question to be asked of these reform processes is whether the newly established oligopolistic structures would be sustainable in the absence of favourable regulatory treatments. Would a full circle be seen in national economies of size less than that of the USA were neither competition law nor special regulation of telecommunications competition to protect the smaller, or the less efficient (perhaps the old monopolist), from being swallowed or co-opted by the great powers of the industry? Presently, it is hard to see any government of an OECD member nation allowing such an outcome, for reasons relating more to the ideology of the market than to strictly observable aspects of economic efficiency. It was certainly the opinion of one respected observer that *de facto* monopoly would be the outcome of unrestricted competition (Phillips, 1985).

Microelectronics and computing begins the convergence process

The technological trajectories which both fostered and were fostered by the breakdown of the natural monopoly rhetoric in telecommunications were those of the computer and microelectronics. It is hard to see how scale economies in infrastructure provision could have been broken down without fully electronic systems of control of signal propagation. And without computers to provide flexible software based control systems the cost of specific hardware solutions would rule out much of current network architecture. The trajectory of technological change has thus been away from highly specific assets comprising the telecommunications network and toward general assets, and thus mass produced not simply for the telecommunications market but for computing markets in general. The specific assets of more significance today are software, the intellectual investments made to bend the general hardware at the core of what the visitor to an exchange sees, to do the signalling and switching which comprises the information flows that are the networks' stock in trade.

It is also this software based technology which creates the multi-layered structure which allows the dis-integration of what had, in days of copper and cross-bar, been a single infrastructure entity. It did not make sense to distinguish carrier from carriage provision until a separable layer or set of layers existed in which the separate services could be distinguished.

Given the fundamental importance of these technological changes, the conventionally accepted view is that without the cost saving improvements in electronics, micro- and opto- included, few of the present or contemplated developments in telecommunications would have been possible. There are two central questions for the agenda. The first is whether that conventional view is supported by the facts of history. The second is to what extent these technological changes were beyond the control of the telecommunications sector, exogenous and imposed by force of technological possibility for cost saving and service innovation. These are both great questions demanding extensive research which will doubtless provoke great controversy.

A mature industry has its second childhood

These two forces for change, the institutional and the technological, combined to throw an industry which had reached a mature stability by the end of the 1960s into the maelstrom from which it is still some little way from emerging. Viewed in a Schumpeterian context we can say that the swarm of innovations, institutional and technological, which broke upon the industry did indeed destroy the comfortable natural monopoly equilibrium which may have had some considerable claims for economic efficiency. But the gains from the turmoil of recent decades are increasing perhaps more rapidly than initially. The telecommunications system is on the section of a sigmoid curve before the inflection point. Markets are still turbulent. New services have yet to achieve their equilibrium configuration for general acceptance by users. Suppliers are still uncertain about their ability to retain market share and revenue share. Firms at different levels in networks of producers serving the sector cannot rest assured of their current places, nor can they ignore possibilities for invading networks served by rivals. The emergence of maturity after this stage of transformation is far away in time. The swarm of innovation continues. The endogeneity of much of that innovation is clear, but demands investigation at both the micro and the mesoeconomic levels.

How to understand these changes and their continuance?

Frameworks within which to understand these changes are being developed well beyond the conventional frameworks appropriate to the study of industries in equilibrium or steady states. The remainder of this paper will examine some of the possible directions for research and how the questions within these programmes relate to each other and to the disequilibrium frameworks under discussion.

4. Fundamental features of the economic nature of telecommunications industries

Demand for telecommunications services, the possibilities for which are not well understood by users or suppliers

Household consumers have no idea of possibilities and are subject to marketing efforts by suppliers who themselves don't know what they are selling. Beyond POTS, services to households define themselves gradually. The answering and fax machines are well understood after some decades of diffusion. The ways in which these are used differ from the initial expectations, ways which are important to marketing of these devices, for example, answering machines are used to filter incoming calls. Further extensions of services to households are similarly lacking in accepted configuration, and will evolve into niches which may or may not be viable to service suppliers. The Minitel experiment in France is the most spectacular example of radical service extension which revealed unsuspected market niches. The implications for market structure are that experimentation and diversity hold uncertain prospects for both private profit and social development, while reluctance to experiment replaces uncertainty with certain stagnation.

The study of demand within the confines of the economist's individualistic framework is severely limited. Marketing and advertising have long recognized the importance of the community of the potential buyer, the relationships between users which create the effects lamely titled "externalities" by economists, the relationships between suppliers and buyers known as "goodwill" of the supplier. As long as the social construction of demand is regarded as a side issue of externalities in demand the explanation of formation of demands for new services and diffusion of usage will remain largely *ad hoc*. Service providers whose minds are captives of economic theories of demand will merely observe "turbulence" and shake their heads and hope for things to settle down. But this turbulence is endemic to markets whose entrepreneurs are searching for profitable avenues to exploit radical technological possibilities. The closest economics comes to dealing with this is via diffusion models. But these models, too, are severely limited. The seemingly inexplicable turbulence of new markets relates to concepts of potential user cognition, uncertainty about the nature of services the new product supplies to the user and to the poorly defined configuration of the product. Understanding these phenomena is a major task. It should not be left to economists.

Commercial demand

Major commercial consumers have demands for services that suppliers struggle to meet. The most important issue for market structures with respect to commercial telecommunications is that of the role of the impersonal market as against that of the alternatives. Major users are so intensive in their use of telecommunications that forms of incomplete integration are clearly being observed. At the lowest level, the telecommunications suppliers create specialist functional departments to manage relations with these customers. At the highest levels these customers integrate backwards to supply their own services in private networks. This is by no means a new phenomenon. Before the existence of a public switched network railways had their own telegraph system. These survived in some cases to become the basis of entry to a competitive telecommunications service market.

The uncertainty about service provision is partly a relationship between expectations raised by technological enthusiasm and actual performance by suppliers. The phenomenon I have labelled "enthusiasm" is inchoate but can be represented by the popular journalism of computers and telecommunications in the context of a business environment focus on rapid communications as a competitive tool, a means of raising productivity. Thus attention of business management is turned to the ways in which telecommunications seems to promise solutions to problems, for example allowing teleworking, reducing the time element in a "just in time" inventory system, promising instant feedback from their customers on quality of their goods and services. But actual telecommunications performance is inevitably lagging these expectations. It cannot be otherwise in this environment of rapid technologically based innovation in telecommunications service provision and products.

Does this mean that major telecommunications users will, for the foreseeable future, be drumming their fingers and frowning at their telecommunications suppliers? Does this mean opportunities for innovative entry to churn existing relationships between suppliers and their major customers, and thus improve the prospects of entrants beyond what might be possible in more stable times? I suspect so. The uncertainty which flows from these considerations may be one of the more important incentives to create forms of integration generally called "networking".

Networks characterize not only the outputs of services provided but also the means of provision

This issue is, of course, the topic of another chapter in this volume. However, it is necessary to make at least the comments below to set the context within which market structures evolve. In fact, it could well be argued that it is a topic generally contiguous with that of networks, as the market structure is one of networks of complementary and competitive firms serving the sector's markets.

Telecommunications services are communications channels which are therefore a complex social network, demand for which is not merely an individual choice. Economic models of externalities treat demand for telecommunications services as a special case, one which includes one or another auxiliary assumptions about the way in which individual utilities are connected. As I pointed out above, the demand for the interconnections which represent the network could well be related, at least for major users, to degrees and varieties of integration. The social and economic construction comprising variously integrated users' demands might well be better seen in the context of theory far removed from utilitarianism. How this links with market structures is an issue to be explored by network theory.

Technologies embodied in service provision encompass a number of once disparate industries, electronics, computer hardware and software, broadcasting. Convergence between information and communications technologies is the result of this set of changes. The meaning of convergence is never all that clear, but perhaps it means membership of a common network or series of alliances of firms whose interests had previously been remote from each other. The most spectacular example of this recently has been the convergence of media broadcasting interests with those of telecommunications service providers. For investment in physical telecommunications networks, there must be demand for the services which may be provided over the physical network. Thus the physical telecommunications network provider must attempt to tie up broadcasting providers which can give a credible chance of sustainable revenues to the physical network.

To what extent is this convergence phenomenon technologically determined? Is it the inevitable outcome of exogenous forces which have fallen on telecommunications, or is the convergence an artefact of strategic decisions by firms with critical roles in the innovation process? And if this is the case, were they taken for or by one or another set of interests in those once disparate industries?

Complexity of resource coordination has led to supersession of open markets in a significant part of the sector, just as open markets were expected to become more significant. Who would have expected that opening telecommunications to competition would result in the impersonal market not becoming the central institution of a decentralized resource allocation mechanism? Well, perhaps a lot of people, especially those with knowledge of other complex contestable oligopolistic production sectors. Open, impersonal, markets do not give the degree of certainty required for sinking investment funds into the creation of productive capacity. An implication is that the formation of networks of alliances by firms in convergent industries becomes one of the more important aspects of business strategy, as firms attempt to gain credible commitments to closely complementary investments, and ensure the same credibility against competitive investments (Richardson, 1972). And as Richardson pointed out at length, the open impersonal market place is no institution for forging such commitments.

Ideological commitments to impersonal markets are commonly held by governments, which insist on competitive tendering procedures to ensure proper accountability to constituents. Sadly, this commitment can cause as many problems as it solves in the highly interlocked environment of rapidly evolving and very high sunk cost technologies. Competitive tendering can sever links essential to continued innovation. In particularly ham handed cases collateral losses may be sustained by loss of the relationship, e.g., writing off obsolescent equipment which could have been traded back; losing use of previous suppliers' services which precipitates obsolescence. Competitive tendering can also prevent the development of long-term relationships, so forestalling the possibility of gains from complementarity which Lazonick (1991) saw as the essential difference between proprietary and managerial capitalism for their productivity performance over the last century.

Thus networks of suppliers, integrated to greater or lesser degree, are emerging to internalize the externalities which otherwise would be an uncontrolled element in the cost and production relations of core suppliers. We observe both (re-)integration and dis-integration in the turbulent telecommunications industries. In the closely related computer industry an extremely clear progress has been observed. The industry quickly evolved into a completely integrated one, where hardware, software, and services to users, were the product of a single firm, whether IBM, DEC, or Fujitsu. This held sway until the 1960s. But from there, disintegration began, with applications software and services to users forming distinct industries. Then following the micro-computer and the workstation, operating system software became somewhat detached from hardware. More recently, some re-integration has occurred, with the formation of a number of important alliances between firms whose interests and expertise were complementary. And in at least the case of

IBM's alliance with Apple, the cause was a strategic choice to attempt to limit the hegemony of the software house, Microsoft, in personal computers. It is of interest that Microsoft, as well as planning to take over the Internet browser market, has recently made noises about investing in telecommunications hardware, in the provision of an elaborate satellite system as the platform for a very large broadband distribution network.

Trajectories of major technologies are at an early stage with great scope for further development

Theoretical possibilities in computing, opto-electronics, microelectronics, wireless, satellite are all so extensive that directions of development are numerous. An example of possibilities is in development of satellite delivery of telecommunications services. Early generations of satellites embodied techniques which severely limited the number of channels. Optic fibre seemed clearly advantaged for long-term investment in multiple channel delivery systems. But that clear advantage is no longer so clear. The uses of satellite now include mobile phone, a use which would have been unthinkable a decade ago.

The size of global demand for capacity to provide telecommunications channels may be so great that many of these possibilities are actually taken up, as in the case of satellite, which could easily have been "locked out", to use that phrase as per Brian Arthur (1989). The clear implication of this argument is that directions of change are not well described as "technologically determined". The great diversity of technological possibility is one of the most important features of the sector. It means that predictions about the future of market structures cannot be based on deduction from "the facts of technology". Choices are being made, some of which do lock in the choosers, but whether they lock out possibilities for competitors is more uncertain.

The smaller the system being considered, the more likely lock in/out will be observed for that system. But points of choice will appear historically even for systems usually locked in to a technology. An example of such a choice node is in the market struggle between the Microsoft MS-DOS operating system and Macintosh: After the Macintosh had gained a substantial share of the market Microsoft decided to compete by adopting one central characteristic of the Macintosh, the graphical user interface, and produced the Windows operating system, essentially an add-on shell to MS-DOS to give it some of the user friendly character of the Macintosh system. There was nothing inevitable about Microsoft's action; alternative strategies were available which may have had at least as good a chance of dealing with the Macintosh threat. But having adopted the GUI environment of Windows, Microsoft and the PC world became locked into a particular trajectory of development. This doubtless pleased Apple shareholders because it implied that the Macintosh system would remain technically superior for a considerable time period while Windows struggled to catch up under technical difficulties (for Windows to achieve the same results as the Mac interface it has to solve technical difficulties in dealing with applications software which differ from each other at a deeper level than Mac applications are allowed. Mac application software developers have an extensive range of high level tools which constrain them to particular pathways, the most obvious being use of the Macintosh "Clipboard": select, cut, copy and paste). Once past that choice node, competition between the two systems could be viewed in terms of a selection process in which market performance or acceptance of the competing technologies, and their champions, will see each evolving into a market niche. In this process relative profitability determines investment capability and ability to engage in incremental technical progress to enhance the products and reduce production costs. The sequence of competitive moves, market successes or failures, is an essentially incremental process of evolutionary selection, such as that seen in the models used by numerous authors in Dosi *et al.*, (1988), Nelson and Winter (1982), Metcalfe (1988), Lane (1993).

So while market structures cannot be described as technologically determined, in the sense that exogenous technical possibilities interact with market size to yield the inevitable, technological choices made at some points in time have the effect of locking a system into a set of pathways of change which are the consequence of incremental choices of firms locked into the broader parameters, choices taken in response to market reactions and to network partners' constraints, encouragements or financial inducements. The choices of the firm are conditioned by their immediate experience, incentive and environment. Market structures can be understood if these processes are understood, and can be extrapolated cautiously forward in time only within increasingly broad confidence limits. Developing rigorous historical models of evolving market structures is a task central to understanding the telecommunications sector during its period of upheaval.

An important implication of this indeterminism is that business decisions, taken for any of the reasons they are taken, are more important to the way technology is used and is developed, than are the exogenous characteristics of technology itself. Alternatives vie for market success, uncertainty is considerable, even for centrally powerful players such as AT&T or any one of the great equipment suppliers. Successful innovations depend on factors which are not in principle predictable by the firm except as goals of firm policy. The success of a technology is therefore dependent upon the sponsor or champion of that technology. That champion's success in propagating its chosen techniques is the essence of the trajectory of development. Whatever certainty there is in technology is thus not mainly exogenous, but is an artefact of the market structures of the sector within which that technological competition is taking place. And the championing of technologies is an important part of the process of uncertainty reduction.

The goals of firm policy are expressions of the firm's conscious will, unlike the random mutations of biological selection. Thus the exploitation of technology is a furthering of growth or profit or other strategic goals. Political pressures exerted by large corporations to divert policy to their own ends, as emphasized by Mancur Olsen (1982), result in actions biased to the interests of those corporations. Adoption of technologies which favour those interests, and is permitted by regulatory or other policy action, entrenches "championed" technologies in trajectories which become relatively locked in. Thus indeterminism at the early stage is again overtaken by a limited range of determinism. So we observe action and transformation more often "when there is money to make". It is the case that historical analysis is the most powerful tool for understanding why it is that one rather than another direction is taken, both technologically and institutionally.

The evolution of the Internet is an important example of technological change which seems at first glance to be quite different from other areas in the telecommunications field. Unlike the growth and development of the telephone and broadcasting networks, the Internet apparently grew on the back of other developments, almost unnoticed for many years. It is a set of services riding on many disparate facilities. It is not owned as an entity by anyone. How did it come about that such a seemingly democratic and anarchic institution has come to enjoy such growth, and become so important? Does it violate all that is written about profit seeking and market orientation of the new telecommunications environment? The contrast between the Internet and commercial database and distribution services is strong. The reasons the telecommunications power structures have not absorbed it are obscure. It is a counter example to the highly concentrated structures of the remainder of telecommunications, as well as to the profit goals of the corporate world. It relied, at least up until the burgeoning of the World Wide Web in the middle of the 1990s, on hardware and other distribution and transmission facilities provided by the remainder of the telecommunications and computer industries. What funds it required were provided largely by defence and educational institutions, sometimes unwittingly through the work effort of staff and graduate students who produced the information which flows through the network. It was a tender flower, in that a small change of institutional policy in the USA could prevent information from getting onto the network (the IMF no longer allows its International Financial Statistics to be accessed by ftp, it sells CD-ROM versions). Were the programmers and other computer centre staff required to charge out their time to someone, some users, much of the traffic would have disappeared. Decisions taken on grounds of cost recovery or enthusiasm for market determination of resource allocation in the medium term would inevitably have shrivelled the pathways which allow the development of long term future niches, at least some of which will ultimately be both privately profitable and socially desirable from the point of view of politically expressed demands. There might never have been a World Wide Web.

Uncertainty reduction by exploitation of network alliances, by user initiation of innovation projects, allows investments which the open market would make impossible

Uncertainty reduction is the firm's goal, explicit or implicit, in coping with this technological and demand environment. The formation of cooperative groups that guarantee each other's investment decisions is one means of achieving this. Other forms of risk sharing include financing options which involve the financier in more than simply the purchase of a debt instrument, or make the customer the financier. Again it is seen that the nature of institutions involved in the production system is shaped in part by that system's demands. The best examples of these attempts at uncertainty reduction are in the failures, in particular where a once successful firm finds itself in some trouble because it has reduced uncertainty by ignoring it. IBM gave away market share, and had to purchase survival at some cost by restructuring and re-inventing the firm. Uncertainty about markets was seen entirely within the mainframe context, and dealt with by IBM's superb customer service. Other forms of computing were ignored for a very long time, with unfortunate results for IBM. How might computer technology have evolved had IBM seen the threat of minicomputers a decade earlier?

Market structures in this environment can then be interpreted as responses not simply to scale of investment as in static theory but as response to uncertainty and means of reducing that uncertainty. While vertical integration has long been associated with new technologies for precisely these reasons the range of solutions now being tried just in the telecommunications sector is far greater than previously seen. The fine layering of the production process, as seen in the engineering categories of telecommunications network infrastructure means that the combinations of possible integrations and disintegrations are numerous.

In order to understand the emergent structures it may be necessary to understand the technologies on which they are based and the technologies which they are championing, the two aspects, exogenous and endogenous, being difficult to disentangle. Modelling emergent technologies is a clear part of the positive heuristic of evolutionary models, evident even in Downie's early model (1958) but explicit in the technological change literature which goes so far as to make technology the unit of analysis rather than the firm (Metcalfe, 1988).

5. Regulation is a central influence on market structure, conduct and performance

National regulation has radically altered the environment of service providers and once protected equipment suppliers

Beginning with Judge Greene's 1982 decision, the climate of national regulation has moved away from monopoly solutions. The questions raised by the progressive change of national regulation in a non-decreasing number of countries include:

- The extent to which these changes were initiated by economic factors, whether these were, and are, inexorable, or did the economic factors follow the regulatory changes?
- If the former, what were those economic factors; if the latter, what were the causes of the initiating changes to regulation?
- To what extent did and does technology develop in directions which take advantage of these regulatory changes; is there a "virtuous" (or is it a vicious) circle of regulatory and technological change leading toward more contestable markets in a manner which might be suggested by Brian Arthur's "lock in" analogy (Arthur, 1989)?

Regulatory change has not only made service and facilities provision contestable in a broad manner, it has cut ties between equipment users and suppliers, and forced these networks into new and usually looser arrangements, as outlined above. The still blunt instruments of regulation have many effects that are unintended, some of which may be consequences of removing certainty of markets from equipment suppliers. Whether this increased uncertainty has stimulated progress in technological change or restricted it to narrower bounds of more certain profitability is a question which may be related to regulation. But has the loosening of ties merely stimulated the development of new ties? Have varieties of consortia emerged to replace the vertical integrations made impossible by regulation? I would suspect so. The diversification of groups once regulated out of service or facility supply into those areas indicates the power of integrative forces discussed by Lazonick (1991) amongst many others. Regulation which is based on the ideology of the contestable market may render these arrangements more complicated, especially for once monopoly providers, but cannot force the market structure into the shape of a mid 19th century commodity market.

International regulation has reinforced the opening to competition by encouraging transformation from bilateral treaty control toward open international markets

The old regimes controlling international telecommunications traffic have not completely disappeared. However, the major revenue routes are now much more open to contestants than are the liner shipping trades. Comparison of these two international modes of communications and transportation (telecommunications transports data for processing, just as ships transport wheat, oil and complex machinery, and considerable volumes of mail) may suggest sunk costs and other entry barriers play a similar role. In the shipping case, capacity may be shifted at will, in telecommunications only some forms of capacity (satellite) are so light footed, and then only within limits. Is freedom of entry sustainable as a basic condition of telecommunications only because sunk costs in cable are a commitment? But how might this work in the present volatile technological world of uncertain obsolescent investments? Major investments in intercontinental cable may require some guarantee against entry by alternative modes which have an average total cost less than the marginal cost of the present investment (though it must be admitted that marginal cost of transmission is very close to zero currently, with measured marginal cost being largely the cost of measurement, or billing cost in other words).

The GATT's Uruguay Round concluded with, *inter alia*, agreement to a range of freedoms in telecommunications services. Introduction of these measures to remove restrictions on trade in telecommunications services may be accompanied by other measures designed by national governments to avoid the consequences of the intended freedom of trade. However, the trajectory of international policy is clearly toward contestable international markets, and to the contesting of national markets by international players. As well as this direction in international policy, the technological directions encouraged by these policies are against effective national restrictions. Broad and narrowcasting beyond national boundaries is now a well established phenomenon using satellite means of distribution. There are already satellite based mobile telephone technologies in place at prices only one order of magnitude greater than cellular mobile telephony. The prospect of national regulation maintaining control over national telecommunications for another decade is as great as such control being maintained over broadcasting a couple of decades ago. Only with internationally binding restrictions on satellite deployment could either unrestricted broad and narrowcasting, or satellite telephony, be thwarted.

Politically driven agendas for competition have given positive encouragement to market entry in the face of real resource disadvantages in some cases

Regulation has imposed restrictions on dominant incumbents over and above those required for general competition policy. The additional purpose has been to foster entry. This has been the case in the USA and the UK, as well as in a number of other jurisdictions in which regulatory measures have been imposed at the level of the firm rather than the industry. Opinion amongst economists as to the desirability of regulation discriminating in this way is divided. Schankerman (1994) is strongly against any discrimination, while Armstrong, Cowan and Vickers (1994:120-124) make a strong argument in favour based on a considerable list of potential social gains. At least some of these potential gains have considerable plausibility in the telecommunications sector, but who has established the existence of any of them? The research will follow the policy.

The agenda for competition is buoyed by belief in the eventual efficacy of competition, such that a form of "winner picking" by artificially lowering entry barriers is sanctioned where many other such interventions are forbidden, for example intervention to promote a local supplier of equipment. The belief is in the nature of the final state of equilibrium market structure holding a number of players, or in the benefits of at least transitory competition in imposing market discipline on incumbents and eventual long term inhabitants of the market. Fashions may change in economic policy-making, and the present set of beliefs may sooner or later come to be seen as deluded. Schankerman's conservative policy guidelines (*op.cit.*) may well be more appropriate than the optimism of encouragement for entry at a direct cost to society in the short run.

Regulation and social policy goals apart from economic efficiency

Social policy tends to be treated by economist policy advisers as a constraint on economic policy goals. Universal service, the old criterion of public telecommunications policy, survives in attenuated form: how can some form of universal service be provided in the new competitive environment? The question is not whether the market structures favoured by economic policy-makers should be implemented or not, but whether the social policy goals are affordable. As a footnote on the emergence of contested markets, universal service, and other goals, are not seen as changing policy but merely as minor irritants for policy to deal with. The coalition of policy-makers and corporations dominates the telecommunications sector and the debates surrounding it.

6. The intellectual climate: have academic scribblers of the present and recent past played an important role in setting the agenda of change in telecommunications?

The issue of economic theory and ideology and its successes in directing the public policy debate is one which is clearly a significant aspect of the determination of market structures. Intellectual opinion has not only swung its weight strongly in the direction of competitive, or contestably oligopolistic, market structures, but has also shaped that debate with its language, categories and dimensions of performance which can be granted attention.

What importance relative to "real" factors of technology and demand does this intellectual climate have? Does it overcome reality and impose outcomes which will prove to be unsustainable? Does it allow social experiments which may or may not prove successful? Does it prevent other social experiments or disrupt existing institutions which are performing in an exemplary manner?

What little analysis of these questions that has been done has not been accorded proper attention, for the obvious reason that attacks on, or merely analysis of, the intellectual *status quo* is evaluated by that intellectual elite in control of the debate. And thus attackers can be destroyed by the easy demonstration that the attack is intellectually wanting. By this means discussion of alternatives is internally stifled, at least for some time. But the cycle of opinion has historically not been as long as a Kondratieff cycle.

Research on the significance of these factors in the trajectories of change in telecommunications will be forthcoming mainly in arrears of the coming swing in opinion. But it is on the agenda for the adventurous scholar wishing to be on the crest of the next wave.

7. What are the research priorities for the explanation of market structures and for the prediction of future evolution of structures?

To conclude, the following list attempts to summarize my suggestions as to the shape of the research agenda for understanding market structures in the telecommunications filière:

Market versus planned (but still private) coordination

- Why has the opening up to competition not led to market relations but to alliances, networks of tied or variously integrated suppliers, and long term relationships which are sustained through periods of seeming uncompetitiveness?
- Where government policy or extreme cost pressures dictates adherence to strict cost criteria for selection of suppliers to service providers, have the consequences been poor performance in the longer run?
- Are the non-market relations in the nature of minimizing transactions costs or of minimizing agency costs? Would measures to reduce agency or transactions costs result in dis-integration and the return to market relations?
- Are the non-market relations in the nature of synergistic gains to coordination of resource exploitation, mainly knowledge resources? Would disintegration be expected as the uncertainty inherent in rapidly changing activities disappeared as the activities became routine?

Volatility in market shares for telecommunications services and the evolution of market structures over time. The Churn or Turbulence Factor

- Can a sufficient regularity in market share evolution be seen in the various relevant markets for telecommunications services to allow the existence of one or more selection processes?
- If not, how can the market mechanism be said to be working to reward the relatively efficient and punish the relatively inefficient, as industrial countries' policies seem to assume?
- How important are the markets which are volatile? Are the "bread and butter" markets for POTS still sufficiently extensive and stable, or is customer loyalty to suppliers of long distance and international telephone and fax services not important?
- Is there a cause and effect relationship between the uncertainties of demand, the volatility of disengaged customers (including major customers) and the uncertainty reduction strategies of integration in networks?

National policy and market structures

- Governments at all levels have their agendas on industry development, including importantly, telecommunications and communications. The trajectory of growth of national carriers, analogy with airlines, in conflict with globalization, is part of national competitive races for competitive advantage and economic prosperity.

- Regulatory policies are in part designed to foster growth of competitive telecommunications, in another part to foster growth of competitive telecommunications users. How is the imposition of regulatory constraints and goals affecting the evolution of structures?
- In particular, is there any giving away of direct productive efficiency in the short run from attempts to introduce new competitors under advantaged conditions?
- Are there other agendas in national policies, for example, to do with the power of the state or of special economic interests? Do governments capture the structures to keep power?
- The architecture of the emerging information infrastructure is central to the nature of its utilization. Will a grander Internet, with freedom of access and analogous with the public roads rather than with private clubs, be the model? Or will Compuserve or Reuters financial services be the model? Implications for market structures that will be imposed by national and international policy are considerable.

Exogenous or policy imposed technologies versus endogenous and integrated technologies

- National policy directions and technology.
- Customer supplier networks and integrated technological development: is integration essential to effective R&D?
- Pre-competitive R&D and competition in implementing innovation: is the oligopolistic structure of service provision effective in delivering innovation?
- Does the innovation imperative favour this structure of service provision, or is it a constraint on effective oligopolistic competition?

Customer locations, national industry policies for attracting major telecommunications users to specific locations, international trade treaty obligations

- Without international agreement will fragmentation disrupt wide area service provision due to short sighted national and local policies?
- International agreements under the control of the major powers, e.g., US influence on the last GATT round on services, impose the internationally powerful networks on everyone regardless of democratic pressures for locally autonomous systems.

- Influence of major users on directions of service development retard or encourage locally autonomous provision by non-integrated firms.

To what extent do customers effectively choose the market structures within which telecommunications services are provided?

- Given the technological openness of service provision, is there evidence of choices being imposed by major users to specifically suit their needs, e.g., for contested supply of services?
- Are major users integrating into the telecommunications services industrial networks? This phenomenon is an example of convergence, though not the only or even the main element in convergence of industries once outside the sector.

Notes

1. This includes expansion and diversification, in other words the sinking of funds required to contest a market.
2. Cournot, Bertrand, cooperative profit maximization, are examples.

References

Ansoff, H.I., 1982, *Understanding and Managing Strategic Change* (North-Holland, Amsterdam).

Antonelli, C., 1994, *The Economics of Localised Technological Change and Industrial Dynamics* (Kluwer, Dordrecht).

Antonelli, C., P. Petit and G. Tahar, 1992, *The Economics of Industrial Modernisation* (Academic Press, London).

Armstrong, M., S. Cowan and J. Vickers, 1994, *Regulatory Reform: Economic Analysis and British Experience* (MIT Press, London).

Arthur, W.B., 1989, Competing technologies, increasing returns and lock-in by historical events, *Economic Journal* 99:116-131.

Australian Government: Industries Assistance Commission, 1984, *Telecommunications and Related Equipment and Parts* (Australian Government Publishing Service, Canberra).

Bornholz, R. and D.S. Evans, 1983, The early history of competition in the telephone industry, in: D.S. Evans, ed., *Breaking Up Bell* (North-Holland, Amsterdam) 7-40.

Brander, J., 1992, Recent developments in industrial organisation, in: *Proceedings of the Conference of Industry Economics* (Bureau of Industry Economics, Canberra) 1-25.

Buckley, P.J. and M. Casson, 1991, *The Future of the Multinational Enterprise* (Macmillan, London).

Dosi, G. *et al.*, eds., 1988, *Technical Change and Economic Theory* (Pinter, London).

Dosi, G., 1988, The nature of the innovative process, in: G. Dosi *et al.*, eds, *Technical Change and Economic Theory* (Pinter, London) 221-238.

Downie, J., 1958, *The Competitive Process* (Duckworth, London).

Eliasson, G., 1991, Modeling the experimentally organised economy, *Journal of Economic Behaviour and Organisation* 16:153-182.

Evans, D.S. and M. Rothschild, 1983, The impact of divestiture on the cost of capital to the Bell system, in: D.S. Evans, ed., *Breaking Up Bell* (North-Holland, Amsterdam) 156-189.

Gowdy, J.M., 1992, Higher selection processes in evolutionary economic change, *Journal of Evolutionary Economics* 2:1-16.

Hannan, M.T. and J. Freeman, 1989, *Organisational Ecology* (Harvard University Press, Cambridge, MA).

Hodgson, G.M., 1993, *Economics and Evolution: Bringing Life Back into Economics* (Polity Press, Oxford).

Kay, N., 1992, Markets, false hierarchies and the evolution of the modern corporation, *Journal of Economic Behaviour and Organisation* 17:315-333.

Laffont, J-J. and J. Tirole, 1993, *A Theory of Incentives in Procurement and Regulation* (MIT Press, Cambridge, MA).

Lane, D.A., 1993, Artificial worlds and economics, *Journal of Evolutionary Economics* 3(2&3):89-108, 177-198.

Langlois, R., 1995, Capabilities and coherence in firms and markets, in: C.A. Montgomery, ed., *Evolutionary and Resource-based Approaches to Strategy* (Kluwer Academic Publishers, Dordrecht).

Lazonick, W., 1991, *Business Organisation and the Myth of the Market Economy* (Cambridge University Press, Cambridge).

Littlechild, S.C., 1979, *Elements of Telecommunications Economics* (P. Perigrinus, Stevenage).

Livingston, 1993, Australia's 19th century communications revolution, *Australian Journal of Communications* 20, 1.

Loasby, B.J., 1991, *Equilibrium and Evolution: an Exploration of Connecting Principles in Economics* (Manchester University Press, Manchester).

Loasby, B.J., 1994, Understanding markets, Working Paper, Turku School of Business.

Marris, R.L., 1964, *The Economic Theory of Managerial Capitalism* (Macmillan, London).

Marshall, A., 1890, *Principles of Economics* (Cambridge University Press, Cambridge).

Metcalfe, J.S., 1988, Evolution and economic change, in: A. Silberston, ed., *Technology and Economic Progress* (Macmillan, London) 54-85.

Mokyr, J., 1990, Punctuated equilibrium and technological progress, *AERPP*, May, 350-354.

Mokyr, J., 1991, *The Lever of Riches* (Oxford University Press, Oxford).

Nelson, R.R. and S. Winter, 1982, *An Evolutionary Theory of Economic Change* (Harvard University Press, Cambridge, MA).

Nightingale, J., 1978, On the definition of market and industry, *Journal of Industrial Economics* 27:31-40.

Olsen, M., 1982, *The Rise and Decline of Nations* (Yale University Press, New Haven).

Penrose, E.T., 1959, *The Theory of the Growth of the Firm* (Basil Blackwell, Oxford).

Perrow, C., 1986, *Complex Organisations: A Critical Essay*, 3rd edn (Random House, New York).

Phillips, A., 1985, The reintegration of telecommunications: an interim view, in: M. Crew, ed., *Analysing the Impact of Regulatory Change in Public Utilities* (Lexington Books, Cambridge, MA) 5-16.

Richardson, G.B., 1990, *Information and Investment*, 2nd edn (Oxford University Press, Oxford).

Rimmer, P.J., 1988, Space and communication, in: R.L. Heathcote, ed., *The Australian Experience* (Longman Cheshire, Melbourne).

Schankerman, M., 1996, Symmetric regulation for competitive telecommunications, *Information Economics and Policy* 8(1)(1 March):3-23.

Scherer, F.M. and D. Ross, 1990, *Industrial Market Structure and Economic Performance*, 3rd edn (Houghton, Mifflin, Boston).

Schumpeter, J.A., 1943, *Capitalism, Socialism and Democracy* (Allen & Unwin, London).

Sharkey, W.W., 1982, *The Theory of Natural Monopoly* (Cambridge University Press, Cambridge).

Silverberg, G. and Verspagen, 1994, Collective learning, innovation and growth in a boundedly rational world, *Journal of Evolutionary Economics* 4:207-226.

Silverberg, G., G. Dosi, and L. Orsenigo, 1988, Innovation, diversity and diffusion: a self organisation model, *Economic Journal* 98:1032-1054.

Snow, M., 1995, The AT&T divestiture: A 10 year retrospective, in: D. Lamberton, ed., *Beyond Competition: The Future of Telecommunications* (Elsevier, Amsterdam) 207-226.

Sutton, J., 1991, *Sunk Costs and Market Structure* (MIT Press, Cambridge MA).

Williamson, O., 1985, *The Economic Institutions of Capitalism* (The Free Press, New York).

Williamson, O., 1993, Transaction cost economics and organisation theory, *Industrial and Corporate Change* 2(2):107-156.

Winter, S., 1991, On coase, competence and the corporation, in: O. Williamson and S. Winter, eds, *The Nature of the Firm* (Oxford University Press, Oxford) 179-195.

CHAPTER 6

Comment: Donald Lamberton

Nightingale has provided an excellent survey of developing ideas about the dynamic aspects of market structure. As he suggests the "old" industrial economics was too simple; but the "new" seems too complex and we can aspire only to special cases. Perhaps it has been just this complexity, with shifting weights attaching to the various, and oftentimes unrecognized, determining factors, that has accounted for both lay and professional redefining of the central concept of competition.

There have been so many different versions: perfect, pure. imperfect, monopolistic, workable, open, potential, sustainable, dynamic, acceptable, constructive, and even fierce and cruel as opposed to "not ungenerous" (Lamberton, 1995a). Much of the logical difficulty that has ensued has arisen from efforts to explain developments stemming from or involving technological and organizational processes of change in terms of equilibrium theory.

There are important consequences for policy, both corporate and governmental. Opportunity is given to gross simplification, dependent upon individual preferences. In contrast to earlier ideas about open competition, as expressed, for example, by Alfred Marshall and P.W.S. Andrews (Lamberton, 1995b), the term has been applied to managed competition (Department of Communications and the Arts, 1994).

A further illustration is the sleight-of-hand by which such managed competition is credited with the benefits (but not the costs) of new technology. We should be asking how much of the technological development of the past century was created and put into operation by non-competitive industrial organization? What do current technological developments owe to achievements prior to the current deregulation and privatization fashion? This, of course, brings us to the role of the historian which is discussed by Ann Moyal in chapter 11.

References

Department of Communications and the Arts, 1994, *Beyond the Duopoly: Australian Telecommunications Policy and Regulation*, Issues Paper, Canberra.

Lamberton, D.M., 1995a, Technology, information and institutions, in: D.M. Lamberton, ed., *Beyond Competition: The Future of Telecommunications* (Elsevier, Amsterdam) 3-10.

Lamberton, D.M., 1995b, Open competition, *Telecommunications Policy* 19(7):587-588.

CHAPTER 7

Telecommunications and economic development*

Rodney Maddock

1. Telecommunications as a causal factor in growth

The existence of a strong positive correlation between the number of telephone lines per capita of a country and some index of its economic well-being is now taken for granted. At the same time we also know that richer countries tend to spend a higher proportion of their national income on telecommunications services. These concepts however tell us very little about the role of telecommunications in economic development. It is quite possible that human beings, being gregarious by nature, like to communicate with one another, and, as they get richer, they indulge this preference. The more interesting element of the link between teledensity and income levels focuses on the possibility that causality runs in the other direction: that greater telecommunications densities lead to increases in incomes.

Formal modelling of the relationship has been carried out using a number of different methodologies. These include the two sector growth model of Park (1988), and models by both Norton (1992) and Karunaratne (1988) of the contribution of telecommunication to growth in a way which is consistent with the recent work in the new growth theory, while Bowles (1994) develops a model which takes into account both supply and demand side factors. These works can be seen as extensions of Perl (1983), Kridel (1987), Taylor and Kriedel (1985) and Bodnar *et al.* (1989) all of which present developments of the simplistic Jipp Curve (1963) which was based on simple regression of teledensity against income.

Recent econometric evidence has tended to suggest that causality is in fact mutual. Lee (1994) finds evidence for this hypothesis using Granger-causality testing on Korean data. The effects there are clear and strong in both directions. Cronin *et al.* (1991) produced similar results for the US over the post-war period in a paper which built on previous work by Hardy (1980).

---

* This paper was developed from discussion held at a telecommunications workshop held in Canberra in July 1994. I would like to thank the Australian Bureau of Transport and Communications Economics, Maurice Haddad, and Don Lamberton for their organization of that conference and participants for their comments.

Belief in the two-way nature of the interaction also lies at the heart of Wellenius (1993) and shows up in well specified models generally. Qualitative research comes to similar conclusions. Mody and Dahlman (1992), for example, show that the rising information content of economic activity worldwide is making the use of information technology increasingly relevant in a broad range of countries and activities. The strength of these cross effects is still open to some debate, with no firm parameter sizes yet accepted, and a later section of this paper deals with the way in which policy choices of governments and operators can have a decisive effect. This switch in emphasis from the fact of feedback from telecommunications investment to growth is highlighting the range of options available to governments and development agencies. Some consideration of this heterogeneity has been revealed by Doran (1989) which shows how middle range countries have tools available which differ from those accessible to the less developed nations. The application she has chosen is interesting, focusing as it does on the role of middle powers in the International Telecommunication Union. Both these studies consider the telecommunications-causes-growth line of argument.

This idea that telecommunications systems increase economic growth rates and promote economic development, provokes a series of questions which motivate this review of the research literature. The next section (Section 2) focuses upon the mechanisms by which the two are linked. It looks mainly at microeconomic, firm level, studies in order to gauge better the precise routes of influence. Section 3 then looks at the aggregate evidence in order to see which countries are doing better than one another in structuring telecommunications investment in a way which enhances growth. The final substantial section (Section 4) considers whether the failures and successes evident in Section 3 can be traced from what we know about the causal mechanisms we have explored in Section 2, or whether there are other significant factors which need to be addressed. This is followed by a brief conclusion.

2. How does telecommunications cause economic development?

Despite the strength of this relationship, the precise mechanisms by which teledensity increases income levels are still poorly understood (cf. Wellenius, 1984). If we retreat to standard models of economic growth four mechanisms appear most worthy of exploration.

2.1 The leading sector effect

The first of the mechanisms is the direct one whereby the telecommunications industry is itself growing rapidly and countries which have a strong telecommunications industry have at least one leading growth sector.

The evidence for telecommunications as a leading sector has been fairly directly established by ITU (1994) amongst others. The evidence presented there finds many countries increasing the share of telecommunications revenue in their GDP by significant amounts over the last decade. There are actually a number of countries for which telecommunications revenue grew by more than three times the rate of the economy more broadly. Table 1 presents some basic data. From the table it seems that even in those countries such as Hong Kong where the network coverage was good and the usage rates initially high, growth has exceeded overall economic growth. The same was true for economies such as China where economic growth rates have exceeded ten per cent for a decade but still telecommunications moves faster. Even poor and sluggish economies in Africa have experienced the same phenomenon. Jussawalla, Okuma and Araki (1989) in their conference volume assimilate a wide range of literature but essentially arrive at the same conclusion about the positive impact of an expanding telecommunications sector.

Table 1. Increase in telecommunications share of GDP, 1982/92

	%		%
Djibouti	346	China	15
Gambia	307	Sri Lanka	196
Kenya	60	Indonesia	205
Zambia	150	Malaysia	200
Senegal	128	Hong Kong	127

Source: ITU (1994), p.4.

2.2 Diffusion/catch-up effect

The second mechanism by which telecommunications influences growth is that better communications speeds the process of diffusion of technologies and thus facilitates catch-up. The role of telecommunications in the diffusion of new ideas has not been much studied but the diffusion of telecommunications itself is a theme picked up in Antonelli (1991). This work focuses on the diffusion of advanced telecommunications, and provides an in-depth analysis of the ways in which developing countries can overcome information asymmetries and their imbalances and so take advantage of opportunities for catching up. In that work he pays special attention to the effects of adoption on the productivity and competitiveness of downstream industries and indirectly on the economy as a whole. The work is particularly interesting in its findings about the ways in which overall investment dynamism is important for diffusion. This is of course a topic much discussed in the economic history literature about growth and in the endogenous growth theory.

Another important study of diffusion in a development context is provided by Hobday (1990). He examines the challenges and difficulties facing third world countries in the field of telecommunications through a case study of Brazil's experience, focusing on the transition from electromechanical to microelectronic technology. He develops a theoretical framework for analysing the dynamics of the diffusion of technology, at the firm, interfirm, sectoral, and macroeconomic levels. His findings are not optimistic in assessing the potential for developing countries to "leapfrog" older forms of technology. Bhalla and Jequier (1988) however present a more optimistic case using India and Brazil for their work. There is an obvious need to extend this work with further case studies and clearly the Brazilian case should be reconsidered.

Crandall and Flamm (1989) consider the process of telecommunications reform and how that itself has been diffused internationally. They consider how the changes brought about by procompetitive policies in the US telecommunications industry have spread to other developed economies.

2.3 Market integration effect

A third possible nexus between telecommunications and growth suggests that better communications leads to better integration of markets and this increases income by increasing market efficiency. Here trade in services, such as tourism or financial services, is an obvious area to explore. Robinson, Sauvant, and Govitrikar (1989) examine the role telecommunications will make to growth of trade in services over the next decade. Their volume surveys the major issues concerning the development and maintenance of appropriate international agreements to utilize adequately the potential for the internationalization of telecommunications and data service exchanges.

By contrast Heng and Low (1990) caution about the possible negative impacts of telecommunications developments on the service sector performance of developing economies. They use input-output methods to assess the contribution of services to Singapore's economic development but note that services has narrowed its contribution to the balance of payments. They are concerned that a well developed telecommunications system can give competitors access to areas of potential growth such as transport, communication and telecommunication services, financial services and tourism. Unless the local economy can produce such services efficiently the benefits of improved telecommunications may be captured by foreigners. A broader view is however taken by Lanvin (1993) who examines the impact of the new technologies of computing and communication on international trade by using fieldwork and case studies by an international group of researchers. It looks primarily at five key service sectors — finance, insurance, software, travel, and accounting and management consulting — covering their respective markets and products, and describes the impact of information technology and telecommunications and data services on the production and delivery of services and the structures of these industries. The conclusions are far more positive than that put forward by Heng and Low, and in this it follows standard trade theory.

The increased possibilities of decentralized development through greater use of telecommunications, with positive external effects, has also been considered. Parker and Hudson (1992) for example describe how rural communities can achieve development goals through state government policies designed to exploit telecommunications potential for rural development. Schuler (1992) too argues that telecommunications developments provide scope for considerable redevelopment of economic regions, and like Parker and Hudson argues for strong policy intervention to achieve this. He contends that while for economic efficiency and technological development purposes it may be important to have private providers of individual components, the efficient planning, coordination and method of recovery of common network costs over the entire system require public oversight. Furthermore, the transportation and telecommunications systems interact, sometimes as substitutes, sometimes as complements, having long-run ramifications for the spatial configuration of economic activity and the long-term use of this infrastructure. While Schmandt, Williams and Wilson (1989) support this same general position, Cook and Beck (1991) are pessimistic on the question of whether the new telecommunications structures will actually enhance decentralization. Here again we have an open research question.

2.4 Better internal management effect

Finally, better telecommunications can also lead to improved management within corporations so that firms make better decisions and plans are more perfectly implemented. This is clearly one of the most difficult areas of research because of the way in which better information impacts on the quality of managerial decision-making. There have been some attempts to use growth accounting methods to measure the contribution of information technology to growth but argues that this method was inadequate because it treats energy and information as equivalent, whereas they are qualitatively quite different.

Antonelli (1989), in his work on the Italian manufacturing sector, finds that telecommunications demand is larger in industries with higher wages, larger sized firms, lower levels of vertical integration and greater product differentiation. Cronin *et al.* (1993) develop a similar model in the US context. Amongst the important case analyses of the microeconomic impact is the work of Eliasson (1990). Like Antonelli, the emphasis is on understanding how information needs are generated within firms and how information is used. This sort of pioneering work at the microeconomic level seems to be the most appropriate way to develop a better understanding of the systemic contribution of communications to firm performance. Hagström (1994) provides another case study of the way in which firms organize their activities, in this case R&D activities, so as to manage their internal activities.

3. Who is winning and who is losing?

The question of why some countries had done poorly in telecommunications development was the focus of the Maitland Commission's deliberations (1984). That report focused attention on the low priority some governments gave telecommunications, on the need to develop commercially viable networks, and on the importance of the foreign exchange restriction to network investment. In the decade since then considerable improvement has been made. Globally, the number of lines available has increased by more than fifty per cent and the less developed countries as a group have increased marginally their share of the world's lines. From Table 2 it is quite clear that there is considerable convergence underway although some fifty countries still have fewer than one main line per 100 inhabitants, that is, growth has been from a low base.

Table 2. Main Lines by Income Group

	Levels	Growth rates 1983-92	
	1992	Main Lines	Main Lines per Head
Low Income	0.80	13.7	11.3
Lower Middle	7.06	8.9	6.8
Upper Middle	7.79	7.9	6.3
High Income	49.14	3.6	3.0

Source: ITU (1994), Table 2.

As well as the obvious differences in teledensity when looked at from the standpoint of income, there are clear regional differences both within and between countries. Of the "low income" countries more than three-quarters have a teledensity in the capital city which is ten times greater than the national figure. In India as an example, the overall rural density is less than one line per 1000 inhabitants compared with four per 100 in the four largest cities; there are at least 200,000 large (greater than 500 inhabitants) villages which do not have telephone access; and possibly 500 million people have no access. In China, the figures are similar. Bowles and Maddock (1994) suggest that half the villages in China still have no telephone service and that is also the case for more than 2000 towns.

Cities then have been winners in the telecommunications development stakes and rural areas have generally lagged. In part this is likely to reflect cost differentials between laying network in congested versus dispersed districts but this hardly seems an adequate explanation. Countries such as India have rural regions as densely populated as the suburban areas of some more developed countries and of course newer wireless based technologies have made the economics of rural networks more favourable. Seen in the light of network cost development differences of around fifty per cent between countries such as India and China, it is not clear that costs *per se* provide an adequate explanation all of the differences. Even if lines do cost three or four times as much in rural districts as in cities, rural lines are likely to generate a much higher proportion of long distance calls so that the cost difference may not reflect the profitability of such expansion. An alternative explanation which needs to be considered is government policy. In the Indian case, a policy of maintaining low prices for rural telephony has had the consequence of diminishing revenues and hence line profitability. This strategy was part of a systematic bias in policy towards rural areas of which low investment is an unintended consequence. In the Chinese case, Zhao and Liu (1994) argue that policy during the eighties was biased towards developing the urban network and backbone long distance network. Rural telephony was a lower priority. Rural areas generally have been comparative losers in the development process.

In regional terms, as in many other areas of development, sub-Saharan Africa has been the worst performer. Information for this region is sketchy but Alleman *et al.* (1994) studies some nine countries in the region. In all cases other than Zimbabwe and South Africa telephone penetration has been growing more slowly than national income and, contrary to most countries, the link between main lines per capita and economic growth is weak although the reverse is not true. While this work is indicative rather than definitive it does reinforce the perception that the poor telephone performance of these African countries arises from deficiencies in their other institutions. This is also the view developed by the ITU (1994). The Arab world shows a similar disparity of experience which reinforces the view that policy determination plays a central role since those countries which have been politically most conservative have lowest telephony development. In Latin America the picture is obscured by the extreme differences in policy stance towards the sector over the past decade. Many of the countries have liberalized their telecommunications sector but at different times and different paces. This has made it hard to compare their situations but Galal *et al.* (1992) finds that liberalization has worked well in the cases of Mexico and Chile. Eastern Europe has followed the Latin American model by relying on liberalization to achieve reform but social and political disruption has meant that the potential gains have not yet been achieved (Holcer, 1994). The Asian Pacific region, again as in many other areas of their economies, has been the most rapidly growing of all regions in teledensity even allowing for income growth. Increasingly liberalization has been used to reach ambitious growth rates but in this case as a tool rather than as the cause of growth.

The use of external telecommunications operators, network managers, network designers, and hardware joint ventures have become essential to meet the rapid network development plans of many countries. Tan (1994) describes the structure of the institutional changes required in China in order to achieve goals of doubling the network size in five years. It is clear that high achieving countries such as Malaysia have consistently invested more than three per cent of national income in the telecommunications infrastructure and that the laggard countries have been investing under two per cent. Revenue reinvestment has been an important source of such funds and this has often required sizeable increases in telephone charges and reconsideration of the dividend operators have had to pay back to the state. As discussed in the Galal *et al.* (1994) study and Crandall and Flamm (1989), the separation of the role of the telecommunications operator from its function as a tax collector, regulator, and welfare institution has been central to the improved performance of operators in developing countries. Again this is not a sufficient condition since both sub-Saharan Africa and the Asian Pacific region invest about one half their revenues in network development.

Multilateral assistance in the telecom area has provided an important additional source of funds to telcos and governments. The major agencies have lent up to six per cent of their available funds for telecommunications in some recent years but gradually seem to be moving away from the area on the conviction that telecommunications projects should be self financing. Bilateral assistance has had some importance but increasingly countries are being expected to make their own arrangements with the private sector in order to meet telecommunications investment needs.

4. Why are some countries doing so badly?

When we look to explanations for the differences in the performance of countries we find that most of the explanations can easily be fitted into a demand and supply framework but that the focus of research to date has been principally on the supply side.

4.1 Demand

Since telephone access in many countries has been restricted on the supply side, it has been natural for research to focus there but demand side differences are equally perplexing. One reads through volumes such as the World Telecommunication Development Report in the hope of finding chapters or even sections devoted to demand with no success, and virtually no discussion of demand factors. Clearly in this telecommunications researchers have been guilty of following an engineering/facilities-based approach and with ignoring the economists' concern about demand. For some developed countries there is a considerable body of research on demand, the classic being Taylor (1994), but little of the work is intertemporal, and little of it comparative (but see Acton and Vogelsang, 1992).

We know little of the impact of quite different pricing systems, such as untimed local calls, or unlimited local calls bundled in with the access price, but introspection suggests that calling habits are likely to be influenced by such differences. Economists are starting to borrow from markets the concept of "growing a market", by which they mean strategically influencing consumer preferences, and in this Cracknell (1994) is one of the few examples.

The impact of social change on telephone usage patterns also appears to have been little studied. As economies enter the market, and as workers drift off to the cities, it seems likely that telephone usage patterns will change but again little work seems to have been undertaken. We know a little about the way in which calling patterns change in some cases as local calling area boundaries are redrawn but again not enough.

The modelling of corporate demand for telephones is far weaker than that of individual demands and I was not able to find a single piece of work exploring this issue in a developing country. Clearly investors are asking assumptions about growth rates in business calling, in the use of leased lines, in data services, in fax etc. in developing countries but there is little work in the public domain. In general we know little about the relative demands for different types of telecommunications media.

4.2 Supply

4.2.1 Market structure

A large literature has emerged suggesting that problems of industry structure are largely responsible for the poor performance of the sector in a number of countries. Examples of this work are Crandall and Flamm (1989), Cowan (1990) and Duch (1991). Snow (1988) has a more optimistic view of state involvement in the telecommunications sector than the other three authors.

Duch (1991) is particularly important for its cross cultural and cross institutional focus, analysing the century-long history of telecommunications policy, focusing on whether and how government ownership actually affects economic performance of firms and on cross-national variations in the implementation of liberalization and privatization policies. In doing this it makes the case that political control, as opposed to state ownership, is the relevant variable explaining economic performance of firms and tests this hypothesis with historical data on the telecommunications sector in over twenty developed nations. This is the position which seems to be emerging strongly from more recent analyses of privatization.

Structural change in terms of privatization, corporatization and competition are all in train in developing countries at the moment. Roth (1987) presented one of the earliest studies of the possibility of the private provision of public services in developing countries. After examining how the public and private sectors differ in supplying their services, he reviews the circumstances that preclude reliance on markets for the efficient supply of needed services. On this basis he argues that what was appropriate in the past may no longer be suitable in view of the impact of new technologies, experiences with regulation and the dynamics of competition, and practical experience in countries as diverse as the United States, China, Brazil, and Yugoslavia. Roth concludes that there is a strong interrelationship among these services, with many of the obstacles to private provision being social and political rather than technical or financial. This remains the dominant current interpretation of developing country experience.

By Crandall and Flamm's criteria Korea would probably be considered an outlier. The Korean model was largely state inspired and has involved many of the features against which those authors warn. Kim, Kim and Yoon (1992) demonstrate that telecommunications in South Korea have grown rapidly under an aggressive government strategy with three key features: massive investment in telecommunications network; alliances among government research institutions, business, and the academic community; and policy measures to foster efficiency in managing the telecommunications sector.

Among the accomplishments are a modern network, the development of Time Division Electronic Switching (TDX) technology, and rapid growth of value-added services. Nevertheless they do caution that the Korean path may not be the one to replicate — changes in technology mean it is now possible to do in a more decentralized way, with greater private participation, what previously had been undertaken through the government.

This theme of the appropriate level of centralization and centralized control has been a major theme of interest in the case of China where decentralized public ownership has characterized many industries but, so far not telecommunications. Davidson, Wang and Hom (1989) argue the case for strong central control but Bowles and Maddock (1994) follow the suggestion of Kim, Kim and Yoon towards a more decentralized approach. Another cautionary tale about reform is contained in Takano (1992) which analyses the case of the Japanese reform. It emphasizes the need for considerable care in managing the interaction of the market structure, market conduct and financial components of the privatization package in a way which is reminiscent of the UK experience. There are a number of studies of privatization and corporatization which have indicated that changes to market structure, or threats of changes to structure, have an important effect on the performance of telecommunications operators. This is clearly an area where we can expect to see a significant number of studies as further experience accumulates.

4.2.2 Managerial failures

Mody and Dahlman (1992) emphasize the importance of management. They contend that effective planning and organizational capabilities are essential to successful adoption of the new technologies. The authors conclude that "managerial skills and entrepreneurship" are key to effective use of telecommunications. In addition to human capital development, diffusion requires a strategy for generating large investments in telecommunications. While proactive measures at accelerating the diffusion have their place, they need to be complemented by reduction of barriers to the international flow of services and investment. This theme is echoed in the work on OECD countries such as Cane (1992). His paper focuses on developed countries, arguing that when used effectively telecommunications and information technology brings gains in the form of better product quality, greater flexibility and speed, and first-mover advantage. He finds that the cost-reducing impact is of secondary importance. He contends that OECD experience has been mixed and often disillusioning. OECD nations, despite their well-developed technological infrastructures, national computer and telecommunications industries and substantial resources, have failed to develop universal rules for applying IT successfully. The obstacles are managerial and regulatory, not technological. Unpreparedness by companies, regulatory barriers and a lack of standards all tend to inhibit successful diffusion.

4.2.3 Sectoral planning

Partos (1989) considers just how small countries can influence the development of their telecommunications sectors. The paper examines the most critical factors of the development pattern and the structure of interests of the Hungarian telecommunications enterprises. It finds that telecommunications restructuring has led to a more competitive manufacturing sector but also one much more deeply integrated into the international manufacturing and interindustry trade. Many firms have not been able to adjust to this environment and restructuring imposes substantial adjustment costs. Partos argues that as a consequence the Hungarian telecommunications industry faces the dilemma over whether to maintain the monopolistic positions (and to lose competitiveness) or to reshape the pattern of development undertaking including gradually domestic and foreign competition. Countrywoman Palocz (1988) was clear, she insisted that reducing the backwardness of service industries in Hungary should be coupled with the opening of the economy toward world markets and that adjustment costs would be small and temporary.

Of course technological choices are also important providing new opportunities for leap-frogging but also providing scope for significant mistakes to be made. Clearly the experience of Brazil is one salutary lesson but the leap towards cellular or satellite provision of services in developing countries where wires have never been used reveals the upside. Goldschmidt (1986) concentrated on the positive possibilities from the use of satellite systems as does the more cautious Hudson (1993).

A separate element of telecommunications policy which can explain some of the differences in the way telecommunications have been developed concerns cultural sensitivity towards media. Many countries have restrictions on the ownership of the media, many have restrictions on content, and some of the same concern has applied to telecommunications. Some governments have preferred that their citizens do not have easy access to the telecommunications media and have consciously restricted development. This explains the inclusion of variables for political freedom which appear in some cross national models as well as explaining why in some countries such as China a disproportionate number of main lines are dedicated to business use (this can of course have other explanations). The effects of central planning can also be to develop the network in ways which do not necessarily accord with the greatest value in use of the medium.

4.2.4 Financial management

Surprisingly little attention has been paid to the question of the financial management of telecommunications corporations and to sectors more broadly. Since telecommunications companies have often been expected to raise internally the revenues they can use for investment, the pricing levels set have an important impact on the quality of the network (Cracknell, 1994). In general the quality of the financial management of telephone systems in developing countries varies enormously but has been little documented. International agencies are however increasingly focusing attention on "institution building" within virtually all infrastructure projects.

5. Conclusions

The pace of expansion of telecommunications systems, and the diversity of new technologies, in developing countries over the last decade has been impressive. This has come about largely as a result of a widespread conviction that telecommunications infrastructure is important to economic development. Despite this our understanding of the mechanisms by which this process operates is still very imperfect. There is clearly considerable scope to understand better the microeconomics of how telecommunications and productivity are linked. There are a number of different views of the macrolevel factors on the supply side which enhance or inhibit the more extensive provision and usage of telecommunications but little understanding of demand side factors. While private firms and funding agencies have tended to focus on supply bottlenecks, there is still considerable scope for further study of the issues both through case studies and through cross national comparisons. With networks expanding so rapidly the opportunities for such research are enormous.

References

Alleman, J., *et al.*, 1994, Telecommunications and economic development: empirical evidence from southern Africa. Paper presented to the International Telecommunications Society Meeting, Sydney.

Antonelli, C., 1989-90, Information technology and the derived demand for telecommunication services in the manufacturing industry, *Information Economics and Policy* 4:45-55.

Antonelli, C., 1991, *The Diffusion of Advanced Telecommunications in Developing Countries*, Organisation for Economic Co-operation and Development, 111.

Bhalla, A., and N. Jequier, 1988, Telecommunications for rural development, in: A. Bhalla and D. James, eds, *New Technologies and Development: Experiences in Technology Blending* (Rienner, Boulder, CO, and London) 269-283.

Boatman, K., 1992, Telecommunications and export performance: an empirical investigation of developing countries, University of Maryland, PhD dissertation.

Bocock, P., 1969, Impact of development: telecommunications in Ethiopia, *Finance and Development* 6:15-21.

Bodnar, J., P. Dilworth and S. Iacono, 1989, Cross sectional analysis of residential telephone subscription in Canada, *Information Economics and Policy* 3:359-738.

Bowles, D., 1995, Telephone penetration: industrial countries vs LDC, in: Donald M. Lamberton (ed.), *Beyond Competition: The Future of Telecommunications* (North-Holland, Amsterdam) 299-314.

Bowles, D. and R. Maddock, 1994, Communications development in China: more haste less speed, paper presented to the Economic Development Conference, Kunming, China.

Cane, A., 1992, Information technology and competitive advantage: lessons from the developed countries, *World Development* 20:1721-1736.

Cook, A. and D. Beck, 1991, Metropolitan dominance versus decentralization in the information age, *Social Science Quarterly* 72:284-298.

Cowan, L., 1990, Privatization in the developing world, *Contributions in Economics and Economic History*, no. 112 (Greenwood, Praeger, Westport, CT, and London).

Cracknell, D., 1994, Growing the telecommunications market — lessons from the past, paper delivered to the Regional Meeting of the International Telecommunications Society, Sydney.

Crandall, R, and K. Flamm, eds, 1989, *Changing the Rules: Technological Change, International Competition, and Regulation in Communications* (Brookings Institution, Washington).

Cronin, F., E. Parker, E. Colleran and M. Gold, 1991, Telecommunications infrastructure and economic growth, *Telecommunications Policy* 5:529-535.

Cronin, F., M. Gold, P. Herbert and S. Lewitzky, 1993, Factor prices, factor substitution, and the relative demand for telecommunications across US industries, *Information Economics and Policy* 1:73-85.

Davidson, W., D. Wang, and S. Hom, 1989, Telecommunications policy and economic development: models for the People's Republic of China, *China Economic Review* 1:93-108.

Doran, J., 1989, *Middle Powers and Technical Multilateralism: The International Telecommunication Union*, Middle Powers in the International System series, no. 4 (North-South Institute, Ottawa).

Duch, R., 1991, *Privatizing the Economy: Telecommunications Policy in Comparative Perspective* (University of Michigan Press, Ann Arbor).

Eliasson, G., *et al.*, 1990, The Knowledge Based Information Economy (Industrial Institute for Economic and Social Research, Stockholm); distributed by Almqvist and Wiksell International.

Folster, S., 1990, The information sector in Sweden: recent developments in firm education, R&D, and telecommunications, in: G. Eliasson, *The Knowledge Based Information Economy,* 89-114.

Galal, A., L. Jones, P. Tandon and I. Vogelsang, 1992, Welfare consequences of selling public enterprises: case studies from Chile, Malaysia, Mexico and the UK, report of a World Bank Conference.

Goldschmidt, D., 1986, The entry of new satellite carriers in international telecommunications: some interests of developing nations, in: D. Demac, ed., *Tracing New Orbits: Cooperation and Competition in Global Satellite Development* (Columbia Studies in Business, Government, and Society, series (Columbia University Press, New York) 88-104.

Gray, C., 1984, The Jakarta telephone connection charge and financing Indonesian telecommunications development, *Bulletin of Indonesian Economic Studies* 139-150.

Hagström, P., 1995, Managing international research and development activities: the role of communications and information technologies, in: Donald M. Lamberton (ed.), *Beyond Competition: The Future of Telecommunications* (North-Holland, Amsterdam) 359-376.

Hardy, A., 1990, The role of the telephone in economic development, *Telecommunications Policy* 4:278-286.

Heng, T., and L. Low, 1990, An Asian NIC's view on service trade liberalization: Singapore's case, *Journal of Economic Development* 15:57-82.

Hobday, M., 1990, *Telecommunications in Developing Countries: The Challenge from Brazil* (Routledge, London and New York).

Holcer, N., 1994, The changing telecommunications environment in CEE, paper delivered at the International Telecommunications Conference, Sydney.

Hudson, H., 1993, Maximizing benefits from new telecommunications technologies: policy challenges for developing countries, in: M Jussawalla, ed., *Global Telecommunications Policies: The Challenge of Change* (Greenwood Press, Westport).

Jipp, A., 1963, Wealth of nations and telephone density, *Telecommunications Journal* 7:52-70.

Jussawalla, M., T. Okuma and T. Araki, eds, 1989, *Information Technology and Global Interdependence*, Greenwood Press under the auspices of the East-West Center, and the Japan Institute of International Affairs.

Karunaratne, N., 1988, Symbiotics of telecommunications, trade and development, *Economia Internazional* 41:44-61.

Kim, C., Y. Kim and C. Yoon, 1992, Korean telecommunications development: achievements and cautionary lessons, *World Development* 20:1829-1841.

Kirn, T., R. Conway and W. Beyers, 1990, Producer services development and the role of telecommunications: a case study in rural Washington, *Growth and Change* 21:33-50.

Lanvin, B., ed., 1993, Trading in a new world order: the impact of telecommunications and data services on international trade in services, *Atwater Series on the World Information Economy*, vol. 3 (Westview Press, Boulder and Oxford).

Lee, C., 1994, The causal relationship between telecommunications investment and economic development in Korea, paper presented to the International Telecommunications Society meeting, Sydney.

Leff, N., 1984, Social benefit-cost analysis and telecommunications investment in developing countries, *Information Economics and Policy*, 217-227.

Leff, N., 1984, Externalities, information costs, and social benefit-cost analysis for economic development: an example from telecommunications, *Economic Development and Cultural Change*, 255-276.

Maitland Commission (Independent Commission for World Wide Telecommunications Development), 1984, *The Missing Link* (ITU, Geneva).

Menon, V., 1989, Telecommunications and development: the case of India, in: M Jussawalla, *et al.*, eds, *Information Technology and Global Interdependence*, 281-288.

Menou, M., ed., 1994, *Measuring the Impact of Information on Development* (International Development Research Centre, Ottawa).

Mody, A. and C. Dahlman, 1992, Performance and potential of information technology: an international perspective, *World Development* 20:1703-1719.

Norton, S., 1992, Transaction costs, telecommunications, and the microeconomics of macroeconomic growth, *Economic Development and Cultural Change* 41:175-196.

Palocz, E., 1988, Services in the world economy, *Acta Oeconomica* 39:171-183.

Park, J., 1988, Growth and telecommunications: a two sector model for telecommunications, University of Hawaii, PhD dissertation.

Parker, E. and H. Hudson, with D. Dillman, S. Sharon and F. Williams, 1992, *Electronic Byways: State Policies for Rural Development through Telecommunications* (Westview Press, London).

Partos, G., 1989, Questions posed by Hungarian telecommunications in the age of microelectronics, *Acta Oeconomica* 40:339-351.

Perl, L., 1983, *Residential Demand for Telephone Service* (National Economic Research Associates, New York).

Robinson, P., K. Sauvant and V. Govitrikar, eds, 1989, *Electronic Highways for World Trade: Issues in Telecommunication and Data Services*, Atwater Series on the World Information Economy (Westview Press, Boulder, CO, and London).

Roth, G., 1987, *The Private Provision of Public Services in Developing Countries*, Oxford University Press for the World Bank.

Schmandt, J., F. Williams and R. Wilson, eds, 1989, *Telecommunications Policy and Economic Development: The New State Role* (Greenwood Press, Praeger, New York and London).

Snow, M., 1988, The state as stopgap: social economy and sustainability of monopoly in the telecommunications sector, *Review of Social Economy* 46:1-23.

Su, S., 1989, Telecommunications and development: the case of China, in: M. Jussawalla, *et al.*, eds, *Information Technology and Global Interdependence*, 289-292.

Takano, Y., 1992, Nippon telegraph and telephone privatization study: experience of Japan and lessons for developing countries, World Bank Discussion Papers, no. 179 (World Bank, Washington, DC).

Tan, Z, 1994, Challenges to the MPT's monopoly, *Telecommunications Policy* 18:174-181.

Taylor, L. and D. Kridel, 1985, Residential demand for access to the telephone network using census tract data, paper presented to the BELLCORE Telecommunications Demand Modelling Conference, New Orleans.

Twenhafel, D., *et al.*, 1989, Telecommunications policy and economic development: the new state role: introduction, in: J. Schmandt, F. Williams and R. Wilson, eds, *Telecommunications Policy and Economic Development: The New State Role* (Greenwood Press, Praeger, New York and London) 1-16.

Wellenius, B., 1984, Telecommunications in developing countries, *Finance and Development,* 33-36.

Wellenius, B. *et al.*, 19--, *Telecommunications: World Bank Experience and Strategy*, World Bank Discussion Papers.

Wenders, J., 1987, *The Economics of Telecommunications: Theory and Policy* (Harper and Row, Ballinger, Cambridge, MA).

Zhao, D. and J. Liu, 1994, Telecommunications development and economic growth in China, *Telecommunications Policy* 18:211-216.

CHAPTER 7

Comment: Nikola Holcer

In relation to the "diffusion/catch-up effect" it may be useful to note the cases where as a result of cheap and good communications on one hand and cheap labour on the other hand, important activities are being shifted out of developed countries to developing countries. Examples includes: (i) in the Caribbean countries many information technology activities (for airlines, insurance companies, etc.) are initiated and successfully sustained giving employment opportunity to many local talents; (ii) outsourcing many software activities to India, and some other countries in Asia from the Californian Silicon Valley; and (iii) production in China of the World Telephone Directory, thanks to cheap labour and cheap communications. There are many other examples in the world.

Hungary and the Czech Republic are exceptions. In the two years 1994 and 1995, partial privatization of Matav and SPT Telecom clearly indicates a trend toward fast development thanks to the strategic partner (more than 15% growth in DEL annually). Similar development models are expected to be developed by the other CEE countries.

On sectoral planning experience, the development in the last few years confirmed the advantages of the joint venture participation or straight buyout by the multinational companies in the manufacturing industries in the CEE countries. Prices achieved in Central Europe are competitive with the world prices for modern equipment.

CHAPTER 7

Comment: David Cracknell

There is a lot of unpublished work going on concerning corporate demand in developing countries, AT&T and BT are both active in India at the moment for example, and some modelling work has been done to create demand projections for corporate as well as domestic telephony.

CHAPTER 8

New developments in telecommunications regulation

Martin Cave

1. Introduction

This chapter presents a personal view of recent developments in telecommunications regulation. Taking a retrospective ten year view, we can see an enormous transformation in the conventional wisdom about how the sector could and should be regulated. Thus at the start of the 1980s, the standard model in OECD countries was for the telecommunications operator (TO) to be a self-regulated government department, typically combining both post and telecommunication functions. There were exceptions. In a small minority of countries, notably in North America, the operator was an investor-owned monopolist, subject to regulation by government agency (although, in the view of some commentators at least, self-regulating to a considerable extent). Other countries were in the process of separating post and telecom-munications functions and "commercializing" the telecommunications operator as a public corporation. Even in such cases, however, pricing and investment decisions remained firmly in government hands.

Contrast this with the situation in 1995. Many operators are now privatized or being prepared for privatization. In the European Union as a requirement, and in many other OECD countries, regulation and operation are now separated. TOs now have much greater freedom in setting places and taking investment decisions, although they are typically still subject to monopoly controls and "social pricing" constraints. Incentive regulation in the form of price caps has to a large extent superseded cost-plus or arbitrary price controls. There have also been massive changes in technology. Fibre is replacing copper as the transmission mechanism for long-distance, and increasingly being inserted in the local loop. Developments in compression are giving even copper greatly expanded capacity, making a reality of the much-vaunted convergence of telephony and audio visual services. One consequence of this is that usage costs are much less distance-dependent than previously. Digitalization has vastly cheapened the switching function and significantly altered the balance of costs favoured access-related or non-traffic sensitive costs. Finally, a whole new range of analog and digital mobile services have become available known by a bewildering variety of initials (DECT, CSM, PCS, etc.).

All of these factors have contributed to the truly fundamental change in the way the sector is perceived — the recognition that much of it is potentially competitive. Chronologically, competition first came to long-distance on a large scale with the growth in the 1970s of competitors to AT&T such as MCI. (Ironically, some observers of the industry, notably Peter Huber (1993) now claim that long-distance rather than the local loop is a natural monopoly.) The UK followed the US's example by licensing a second long-distance carrier, Mercury, in 1983. However, the major development in services arose from the emergence of mobile operators. It now seems clear that the regulatory framework and technological preferences which characterized the monopoly period held back the development of mobile technologies. Initially, mobile telephony was seen as complementary to fixed link services, and often entrusted to the incumbent. However, as the 1980s progressed, two propositions subversive of the monopoly orthodoxy came into prominence. First, it was recognized that the costs of duplicating mobile networks were comparatively small, and that they were natural oligopolies rather than natural monopolies. Mobile concessions were also potentially highly profitable, and governments started to regard the auctioning of licences to rival operators as a significant source of revenue. Secondly, it became apparent that, in some environments, the cheaper mobile technologies were quite capable of providing basic telephony at lower costs than a fixed link network built from scratch, or even an existing one requiring modernization. Thus it is now standard practice in areas where the telecommunications network is under-developed (for example, in much of central and eastern Europe and the former Soviet Union) to view wireless technologies as direct competitors with wire-based ones (Davies *et al.*, 1995). Even in markets characterized by developed telecommunications systems, some wireless operators are positioning their products in direct competition with the fixed link operators. (In the United Kingdom, the PCN service Mercury One-2-One and the projected fixed radio company Ionica fall into this category.)

The final barricade recently breached by proponents of competitive telecommunications is the local loop. Where competition is permitted, it is emerging in this sector from three principal sources: the wireless technologies noted above, which are particularly cost effective in more remote environments, and in many cases provide mobility as an additional service characteristic; fibre networks built in metropolitan areas and aimed particularly at business customers; and entrants from other network utility sectors, which in many cases already have a large customer base and potentially enjoy major economies of scope through joint provision of telecommunications and, for example, cable television or electricity distribution.

Worldwide, the forces favouring the removal of restrictions on competition now look unstoppable in the European Union. Member governments have formally agreed to introduce both services and infrastructure competition from 1998, with modest derogations for certain less well developed member countries. Other countries too are opening up their markets. Australia has announced the end of its current duopoly policy. Most notably, legislative and regulatory developments are opening up the US market.

These developments are both the product of and impose new requirements upon regulation. In this chapter I review some of the problems raised and responses to them found either in the practice of regulatory policy or in regulatory proposals. Section 2 examines some of the issues concerning the institutional framework of regulation. Section 3 reviews certain key regulation issues of current concern while Section 4 discusses the future regulatory agenda.

2. The institutional framework of regulation

Tolstoy noted in Anna Karenina that while all happy families are similar, each unhappy family is unhappy in different ways. To judge from the diversity of institutional arrangements for telecoms regulation, no country is entirely happy. The options vary from:

(a) Self-regulation by the operator, there the latter is normally part of a government department and the minister is the regulator.
(b) Regulation by a government department of the operator, in the form of a public corporation. (Note the possibility of a conflict of interest between the department as owner and the department as regulator.)
(c) Regulation of an investor-owned sector by a government department.
(d) Regulation of a publicly or privately owned sector by an independent regulatory agency, responsible to either or both of the government and the legislature.
(e) No special regulation, beyond general competition law.

There are, of course, numerous versions of each. Many operators are subject to the double jeopardy of an industry-specific regulator and a competition authority. In many jurisdictions, there is an appeals mechanism involving either a "higher" regulatory body (the Monopolies and Mergers Commission in the United Kingdom for example) or — more significantly — the courts. Thus it can very plausibly be argued that a principal regulator of the US telecommunications sector since 1982 has not been the FCC and state regulatory bodies, but Judge Greene of the Federal Appeals Court, implementing the modified final judgement which achieved the divestiture of AT&T.

We also observe with increasing frequency the operation of multi-level or overlapping regulation within a (nation) state-federal system. The co-existence of two-tier regulation has been a feature of US regulation since its inception. Increasingly, the European Commission is playing an upper-tier role within the European Union. The advantages and disadvantages of these multi-tier systems require urgent research.

It is, of course, possible to identify a range of objectives for and constraints on the regulatory system from a political or administrative perspective. In a democracy, there should be some form of legislative mandate and an appropriate level of democratic accountability. The rights of individuals or shareholders should not be over-ridden by this process. However, from a narrower perspective of economic efficiency, the requirements for efficient regulation derive from essential characteristics of the cost base of network utilities.

Their production processes normally exhibit two characteristics they contain elements which are natural monopolies, and they are characterized by long-term sunk investments. The natural monopoly property can itself be defined in one of two ways. The strong definition requires decreasing average costs over the whole range of output. The weak definition requires that the market is most cheaply served by a single "efficient" supplier. Many network-based industries are likely to satisfy the strong definition in terms of numbers of customers served, as hooking up an extra customer often imposes low marginal costs. But this leaves open the question of whether the marginal cost of providing more units of output to a given set of customers either rises or falls. It also leaves open the question whether the supply of a service (i.e. the transaction covered by a contract between supplier and customers) is just as much a natural monopoly as its transportation or delivery. In parts of the telecommunications industry, for example, it may be the case that while transportation and storage are natural monopolies, services are a potentially competitive activity.

While there is considerable controversy over the extent to which any particular segment exhibits natural monopoly properties, it is widely accepted that, where natural monopoly exists, some kind of control over prices is necessary to prevent the exploitation of customers and to ensure that the incumbent monopolist operates efficiently. Thus regulation is necessary to protect consumers. At the same time, some limits on regulatory discretion may be needed to ensure continuing supply of the good or service in question, as a result of the long-term and sunk nature of much investment in utilities. The problem here is that, unless there is some restraint on price controls investors in utilities may be subject to expropriation through the regulatory system. Imagine, for example, that a price control is imposed which allows recovery of running costs, but prevents even an efficiently run company from earning a return on network investment equal to its cost of capital. Even though the firm would not close down at once, the implementation of such a policy would clearly make it impossible for it to attract new funds. Consumers might benefit in the short-run as shareholders lost their investments, but the industry would gradually wind down. The possibility of such an eventuality may by itself be enough to depress share prices and stifle investment.

Viewed in this light, the regulatory system needs both to protect consumers from excessive prices and to protect investors from expropriation of the value of their sunk investment. The role of the regulator is thus to act as an intermediary, both controlling prices and assuring the supplier an adequate rate of return. The regulator thus concludes, on behalf of consumers, a "contract" with the regulated firm.

The framework governing that contract varies from country to country. In the United States, state or federal regulatory authorities are subject to appeal to the Courts which are heavily involved in regulatory policy. Thus in a series of landmark judgements, notably the Hope Natural Gas case of 1944, the Supreme Court laid down principles governing the determination of just and reasonable rates of return (Kahn, 1988, 1:45-57). The principles of regulation are set out in legislation and subject to appeal to the Courts.

In keeping with its traditions, the United Kingdom has adopted an administrative based rather than court based approach to devising a regulatory system which seeks to reconcile the interests of consumers and investors in each industry. The legislative framework provided by the UK Telecommunications Act of 1984 is fairly general with more flesh being put on the arrangements through the more detailed provisions of the licence for each operator. The amount of detail, especially the dominant operator, BT, involved has varied over time subject to a clear trend observable over time for the provisions to become more complex and specific.

Does this provide the required comfort to both investors and consumers? On one hand, it is clear that the regulator/regulatee relationship has little in common with a contract governing a discrete transaction between two private parties — for example, for the sale and purchase of a clearly defined good or service. On the other hand, several writers (such as Goldberg (1976) and Williamson (1985)) have pointed out that many contracts between private parties are long-term and incomplete. They have christened such arrangements "relational contracts" and analysed the circumstances encouraging their formation and implementation. Clearly this analysis requires some adaptation if it is to be applied to the regulator/regulatee relationship, in which the two parties are not on the same "level" as the two parties to an ordinary commercial contract would be. However, it can be argued that the contracting analogy is legitimate because, while the regulator brings to the arrangement certain powers made available by the legislation, the regulated firm typically has the advantage of asymmetrically held information: its knowledge of the technological and market environment is likely to place it at a considerable advantage over the regulator. Thus the formal advantage of the regulator is compensated for to some degree by the informational advantage of the regulated firm.

If it is accepted that it is useful to regard the relationship as a contract or bargain, the immediate question arises of how the arrangement is enforced. From the standpoint of the firm, enforcement in the UK, for example, is ensured by the relevant Act and the licence to which the firm is subject. This will lay down the powers and obligations of the regulator and the firm; in extreme cases, for example, the regulator may be entitled to rescind the licence and prevent the firm from operating. The enforcement mechanism imposed upon the regulator is a more complex one, about which two views can he held. According to the first view, decisions of the industry regulator are subject to appeal of the Monopolies and Mergers Commission on the basis of the mechanisms described above. (They are also subject to judicial review, but regulated firms have not hitherto made use of this avenue to a significant extent.) However, it is probably more natural to view the industry regulator and the MMC as part of the same regulatory system and this raises the question of what incentives there are on the regulatory side to abide by the implicit regulatory contract.

The principal incentive enforcing the regulator to perform the contract is the interest of the customers of the regulated firm. The enforcement mechanism works as follows: the cost of capital to the regulated firm depends upon investors' view of the risks entailed in their actual or potential investment. Breaches of the contract increase perceived regulatory risk and hence the cost of capital and because the regulatory body is obliged to ensure that the regulated company can maintain itself in business it is deterred from breaching the contract by concern for the resulting price rises which would be imposed upon customers.

It should be noted that this proposition about the existence of a self-enforcing mechanism is vulnerable on a number of counts. First, the regulator may be pursuing interests different from those of the customers of regulated firms; second, the cost of capital to a regulated firm may be affected by the

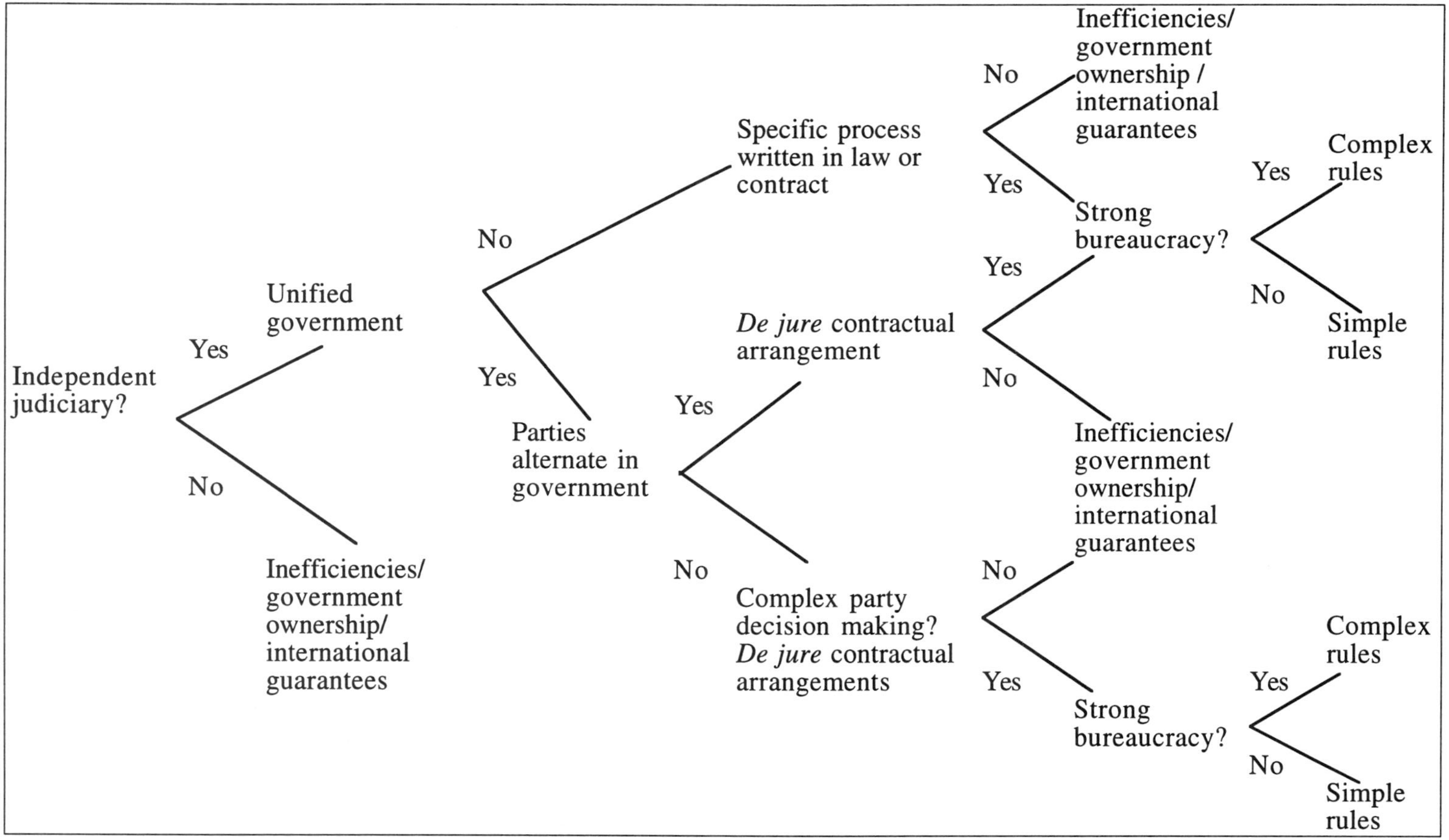

Figure 1. A decision tree model of a country's institutional endowments.

actions of any of the regulators involved in the process many of which will be outside the control of each individual regulator; third, the regulator or investors may not understand the implicit enforcement mechanism. Whilst these arguments suggest that the mechanism may not be fully effective they do not suggest that there is no enforcement mechanism in operation at all.

A particularly rich analysis of the problems of regulatory commitment can be found in Levy and Spiller's recent comparative analysis of telecommunications regulation (Levy and Spiller, 1994). Figure 1 sets out their model of a country's institutional endowment, or its set of legislative, political, judicial and administrative endowments A first key issue is whether there exist domestic institutions capable of deterring arbitrary administrative action; they argue that an independent judiciary is necessary for this purpose. Given an independent judiciary, the predictability of legislation becomes an issue. If government alternates between parties with widely different views, current legislative arrangements provide no certainty for the future. This tendency may, however, be qualified by the existence of overlapping legislative and other authority. Commitment can also be demonstrated by embedding regulatory arrangements in contract law, rather than public or administrative law.

The final distinction exhibited in the diagram is between countries with capable administrations and those without. A strong bureaucracy can make complex rules work in a predictable way.

Encouragingly (and contrary to Tolstoy's proposition above), the authors conclude that a wide range of regulatory procedures can restrain arbitrary administrative action, and hence make the regulatory contract credible. If, however, none of the methods for maintaining regulatory commitment exists, the alternatives may be to accept the resulting inefficiencies associated with private ownership, to impose government ownership, or to rely upon some form of international guarantee.

One particular form of regulatory institution noted above is to rely solely upon general competition law. As is well known, this approach has been adopted by the New Zealand government in relation not only to the telecommunications sector but to other utilities such as energy. In 1989, New Zealand became the first member of the OECD to introduce full competition to all sectors in the telecommunications industry. Recognizing the disadvantages of extensive regulatory intervention, the Government opted for a regime which applied to all utilities, involved "light handed" regulation of the incumbent and relied for enforcement upon the country's generic competition law body, and upon private legal actions.

The New Zealand experience has attracted considerable attention, in part as a result of a protracted legal dispute concerning the terms on which entrants should have access to the network of the dominant operator, the Telecom Corporation of New Zealand. In 1995, New Zealand officials published a discussion paper which invited views on the desirability of relying upon this regulatory framework for defining the regime for interconnection negotiation (New Zealand Government, 1995). One alternative mooted in the paper was the institution of a regime for mandatory arbitration.

The result of the consultation on the access regime is not yet known. It is notable, however, that few other countries have chosen to follow the New Zealand route. Instead, they have generally opted for overlapping regulation undertaken by an industry-specific (but not necessarily independent) regulator and a competition authority. The former typically proscribes certain courses of conduct *ex ante*, while the latter deals with complaints *ex post*. Several regulators have argued that the need for prescriptive powers arises in part from the acute difficulties of regulating the transition from a monopoly to a competitive market. In order to invest in competing infrastructures entrants require a degree of foreknowledge of regulatory arrangements, which is often unavailable from a country's jurisprudence, because competition in network industries is a new development for which precedents do not exists. (However, this begs the question of whether regulators do in fact provide a stable and predictable regime, rather than adapted constantly to new circumstances.)

It is thus probably accurate to characterize the current trend in regulatory institutions as one towards a regime in which (increasingly independent) industry-specific regulators discharge the main regulatory function, often overlapping with a generic competition authority, but increasingly view their task as one of implementing competition law and policy approaches. This involves in particular confining regulation to dominant operators and devoting considerable resources to policing conventional anti-competitive practices such as cross subsidization, predation and abusive vertical relationships.

3. Key regulatory issues

It is not possible in this chapter to provide a review of all, or indeed, of most, of the regulatory issues with which governments and regulators are currently engaged, as they undertake the transition from monopolistic to more or less competitive markets. However, four of them have acquired particular prominence in the discussion, and this section makes some brief observations about each of these. They are: licensing (how many operators to admit to various network and services markets); interconnection (the terms and conditions with which one operator has access to another's network); the issue of how to combine maintenance of universal service obligations with development of competition; and methods of price control. Much of the discussion is based upon examples drawn from the European Union. However, similar issues are arising in all other jurisdictions where competition is being introduced.

A. Licensing

One of the first issues a government has to address is whether and how to control the number of firms operating in particular segments of the industry. In mobile, this decision is constrained to some extent by the availability of spectrum — although governments which really want to increase the number of mobile operators could probably find the spectrum from somewhere if they chose. At the level of service providers there is no obvious reason for any kind of limitation although their terms of access to the network become critical (see below). The fundamental licensing issue is thus likely to become the restrictions if any placed on the number of fixed link operators with their own infrastructures.

The arguments in favour of some kind of restriction typically run like this. If everyone is allowed in, the outcome would be either too favourable, or alternatively too adverse, for the incumbent. It might be too favourable for the incumbent if the entrants dissipate their forces in fighting one another for a limited market share rather than present a serious unified challenge to the incumbent. Conversely, it might be too adverse if the incumbent, attacked from every quarter, and subject to constraints on its tariffing, were unable to survive al the level required to sustain its universal service obligation. A subsidiary but nonetheless important argument, is that granting indiscriminate rights to lay down networks imposes social and environmental costs, which can be limited by selective envy, possibly accompanied by sharing facilities such as ducts or poles.

It is impossible to say *a priori* whether either of the first set of arguments relating to the competitive process, justifies restricted entry. One approach would be to analyse outcomes in the case where entry has been restricted. But here different countries have experienced different outcomes. In the United Kingdom, for example, the period of restricted entry under the Duopoly Policy (1983-1991) brought rather limited effective competition (Cave, 1995). Mercury initially concentrated upon certain markets, especially international calls, made profitable by regulation of the tariff structure. Perhaps under regulatory guidance or perceived guidance, it also entered a range of other markets in an ineffective way. Its pricing strategy mirrored BT's exactly, maintaining a constant discount focused upon large business customers. BT was broadly content, in a fast growing market, to cede the relatively small amounts of market share it lost, even if the markets in question were profitable.

Things changed radically in 1991, for two reasons. Firstly, BT was allowed by OFTEL to offer substantial quantity discounts to large business customers. When they did so, Mercury was unable to maintain the price differential. Secondly, the liberalization of the market unleashed other operators onto the profitable cream which Mercury had formerly been skimming on its own.

Mercury's financial problems since 1994, when it reduced its labour force by one quarter, and the decline in BT's market share since liberalization may suggest that a regulated duopoly only provides an attenuated form of competition. But this may not be so. In Australia, the duopoly has unleashed much more aggressive competition between Telecom Australia and Optus. Aggressive price cutting and customer chasing have been observed, especially associated with the ballot for customers under the equal access policy. The entrant's market share has grown much more quickly than did Mercury's (BTCE, 1995).

In trying to interpret these experiences and their lessons, a natural starting point to adopt is that the burden of proof in licensing policy falls upon those seeking to restrict entry rather than upon those favouring liberalization. It is relatively simple to construct formal economic models in which welfare is enhanced by limiting the number of firms operating in a market by regulation (Armstrong, Cowan and Vickers, 1994). But these normally require rather unusual assumptions, and moreover they neglect the crucial consideration that a process of unrestricted competition is much more likely to generate information about what customers want, through innovation (successful or unsuccessful) and niche entry and exit.

A further consideration is relevant. One way of visualizing the process of entry into a market which was previously an exclusive one is as a race between the incumbent's capacity to adapt itself to the new market conditions (notably the existence of consumer choice) and the entrant's ability to accumulate the necessary assets to survive in a market which by definition must be new to it. The anticipation effect of 1998 is already making incumbent operators adjust to protect their powerful positions in the market. But the form of entry is also changing. In most of the world, the archetypical potential entrant is not a national company setting up a stand-alone business but a joint venture between an overseas telecommunications operator, often from a liberalized market and a major domestic company which already possesses many of the assets necessary for success, in the form of either (or better still, both) an existing private network and an existing customer base. Such entrants not only enjoy economies of scope, but also have extremely deep pockets. Neither they nor the incumbents really need protection, provided the playing field is not tilted by regulation.

Thus the case for restricting entry has not been made convincingly in general terms, although special circumstances may justify such restrictions. Conversely, I am not convinced of the necessity for special measures to promote entry, such as line of business restrictions on the incumbent, However, this does not preclude differential treatment of operators in accordance with the market power which they wield, in such matters as interconnection arrangements. Such asymmetric regulation is an application of a standard competition policy approach.

B. Interconnection

The last year or so has given regulators and operators throughout the world, a crash course in the esoteric subject of the regulation and pricing of access. Numerous industry and academic papers and conferences have been devoted to the subject. This is hardly surprising given that a change of a few per cent in interconnection charges can make the difference between profit and loss for an entrant, half of whose revenues may at the outset go to the incumbent operator.

Fortunately, certain basic regulatory principles relating to interconnect are increasingly being accepted, and the intellectual basis, and commercial consequences, of the unresolved issues are now better appreciated. (See especially Baumol and Sidak, 1993; WIK, 1994; Laffont and Tirole, 1994 and references listed therein; and OFTEL, 1995a).

Thus there is little dispute about the need for an "any to any" principle any subscriber or any network should have access to any other subscriber on any other network. Equally, it is generally accepted that an operator should not provide lower quality interconnection services to other operators than to themselves. However, disputes still prevail concerning the degree of unbundling of interconnection services, the basis for pricing such services and the need for asymmetric regulation of the incumbent and others in relation to interconnection.

On the unbundling issue, the natural presumption is that other operators should be protected from "full line forcing" by a dominant network: in other words that unbundling to meet customers' wishes should, within reason, be required. The general presumption in favour of unbundling may have to be qualified on the grounds of transaction costs or problems associated with incentives to innovate: if the dominant network immediately has to make available to its competitors, at cost, any network facility, it introduces technological advance may be slowed down. But these are exceptions to the underlying rule.

The second controversial issue concerns pricing principles. The options have been widely rehearsed in the form of variations on the following themes:

- Incremental costs.
- Fully allocated costs.
- Ramsey prices (i.e. differentiated interconnection charges aimed at promoting a structure of final services prices which satisfies economic efficiency, in the sense that common network costs are recovered primarily from services demand for which is relatively unresponsive to price).
- The efficient component pricing rule or ECPR (which makes interconnection charging equal to the sum of incremental costs and the opportunity cost to the incumbent of losing the business to an entrant).

The variations on these basic themes include the following:

- Proposals that interconnection prices should be set by the national regulator within a range bounded from below by incremental cost and from above by (what is effectively) forward looking fully allocated costs.
- Adjustments to the ECPR to eliminate from the opportunity cost excess profits made on network operations and costs inefficiently incurred in the generation of interconnection services.

Which of these rules the regulator will choose should clearly depend upon the regulator's objectives and also upon practical considerations: for example, application by the regulator of Ramsey pricing requires a degree of knowledge about demand elasticities which may not be present. However, as Laffont and Tirole (1994) have shown the task of efficient pricing can be delegated to the firm by a suitable form of incentive (price cap) regulation.

The range of possible objectives include the following:

- Ensuring the incumbent's viability (for example, its ability to satisfy universal service obligations).
- Maintenance of an inherited structure of retail prices.
- Prevention of excess profits.
- Deterrence of uneconomic entry.
- Promotion of infrastructure competition.

Regulatory policy on access pricing will have a major influence on the future financial performance of incumbents. In particular, the adoption of interconnect prices based solely on incremental costs might lead to a significant reduction in profitability and market value of the incumbent operator, unless the effects could be offset by improvements in efficiency. Conversely, the efficient component pricing rule would be expected to maintain the profitability of the operator, even as it loses substantial market share. A combination of fully allocated cost interconnect charges with current imbalances in tariffs would be likely to promote entry into profitable markets, deprive the incumbent of the source of their cross subsidies and endanger profitability.

Clearly the task or selecting an appropriate interconnection policy is much simplified by any measures to make retail tariffs more cost-based. At present most countries have arrangements in which the provision of access generates losses, which are cross-subsidized by excessive profits on calls. Equally, a policy of geographically averaged pricing makes some subscribers profitable and others unprofitable. If regulators were content to allow the rebalancing and deveraging of prices, many of the difficulties of interconnection policy would be overcome. In practice, regulators are increasingly addressing the issue of rebalancing by encouraging operators to make available a range of tariffs, which limit the extent of cross-subsidies from calls to access to a limited number of subscribers making relatively little use of the network. Equally, radio-based technologies are reducing the differential in the cost of serving customers located in different regions or environments. Although departures from economically efficient tariffs can still be discerned in most countries, they are still diminishing.

The issue of symmetry in the regulation of interconnect is also important. The UK experience has convinced OFTEL of the necessity for imposing accounting, separation and non-discrimination upon BT, the dominant operator and this policy is also proposed by the European Commission. Separate accounting of network activities provides in principle at least — a method of determining whether network operations are priced excessively. Non-discrimination ensures that network services are "sold" to the operator's affiliated retail organization at the same transfer price as they are made available to competing operators through interconnect prices. Many incumbents seem to regard this as a reasonable approach, perhaps (as they may feel) preferring it to the alternatives of structural or ownership separation, perhaps (as entrants may fear) because they are confident they can manipulate the system to their own advantage. Accounting separation is certainly worth trying, and more radical forms of separation should perhaps only be considered if it fails. However, there is no reason that such provision should apply to all operators, however small. It is more satisfactory to operate some kind of rebuttable market share condition to determine whether there is a presumption in favour of a requirement for accounting separation or the absence of such a presumption. This is probably a reasonable approach.

C. Universal service obligations

The third issue concerns universal service obligations (USOs) in a competitive environment. USOs in telecommunications typically require incumbent operators to provide, on demand, "basic" telephony services to any (sometimes residential) customer who requires it, at an acceptable or affordable price. Because the technology of telecommunications is changing rapidly and because costs and the pattern of demand change too, services covered by the obligations are likely to change over time (see Cave *et al.*, 1994).

In practice there are at least four different policy perspectives that relate to universal service. These are:

- The achievement of universal geographical coverage, requiring that the network or networks are rolled out over the whole of the territory in question.
- The geographic averaging of tariffs. This requires that customers in the same category (for example, domestic or business customers), are charged the same tariffs, irrespective of their individual costs of service.
- Low access charges for residential customers, implying that revenues realized from monthly or quarterly rentals are low relative to the cost of service.
- Targeted subsidies which are typically directed at the poor ,the elderly, the disabled, rural dwellers or low users of the telephone.

The first two of these elements are normally present in every country, supplemented by either or both of the second two.

At present universal service obligations are largely confined to voice telephony, or basic service, and there are difficulties in extending the universal service obligation to incorporate additional services. Firstly, many such services are not fully tested in the market, and it is not certain that households or businesses will require them on an appropriate scale. Secondly, their use is often confined to more affluent customers, and requiring them to be made available at a low price would require less affluent customers to cover the cost of services which they do not want. Thirdly, the extension of universal service obligations increases the net cost of such obligations. If such costs are shared, they may deter competition.

Thus although it is proper that universal service obligations be reviewed in the light of changing economic, social and technological conditions, the case for the extension of the obligations is at present weak. The Commission's Green Paper on infrastructure liberalization concludes that it is premature to extend the notion of definition of infrastructure to new areas.

The cost of meeting universal service obligations in any locality is increased by the sum of losses incurred by operators in serving those customers whom they are obliged to serve under the universal service obligation but whom they would not otherwise serve. In principle, the calculation should be made on the basis of the costs of an efficient operation.

Estimation of the cost of USOs thus involves examination of the costs and revenues associated with particular customers or groups of customers. Since only a minority of customers is likely to impose net USO costs, it is reasonable to estimate the costs of provision to "loss making" customers on an avoidable basis — i.e. to calculate what costs would be saved if those customers and those alone — were detached from the network. On the revenue side, it is natural to treat all revenues paid by a customer who might be imposing a net USO cost as the revenue associated with that customer. However, there is also an argument for attributing some of the revenue of incoming calls to the customer's account, on the ground that that revenue too (or some of it at least) would be lost if the customer left the network. A calculation of this kind has been carried out in a number of countries, and experience demonstrates the practicability of the method.

Existing studies include:

- The path-breaking study of the cost of Australian universal service obligations carried out by the Bureau of Transport and Communications Economics in Canberra (BTCE, 1989). This adopted an avoidable cost approach, incorporated incoming call revenues and ignored replacement effects. The estimates amounted to A$210 million. Recent recomputations have produced substantially lower figures. An estimate of the cost of USOs, made on a fully distributed cost basis at the same time as the original BTCE work was about three times that made on an avoidable costs basis.
- In 1994, the Cambridge consulting firm *Analysis* produced estimates of the costs of USOs throughout the European Union on an avoidable cost basis. They fell in the range of .5% to 3% of the carrier's revenues, except for Spain (5%), Portugal (8%) and Greece (15%).
- In the United Kingdom, OFTEL has produced an estimate of the cost of USOs. Including both unprofitable low revenue customers and high cost areas, and the net cost of other services such as unprofitable pay phones and services for people with disabilities, the estimate fell in the £58 to £89 million range (OFTEL, 1995b).

It is possible, however, that all of these studies over-estimate the costs of USOs. There are a number of reasons suggesting that operators may benefit from carrying USOs. For example, there may be life-cycle effects: customers who are currently unprofitable may shortly become so, especially as tariffs are rebalanced. In addition, operators may derive marketing and other benefits from providing a universal service. These considerations are difficult to value precisely.

Historically, USOs have been funded through cross-subsidy by the monopoly operator. In some cases, further funding has come from social security payments to disadvantaged individuals.

As competition develops, the need for an explicit form of funding grows. Although funding out of general tax revenue remains theoretically possible, most governments regard the sector as a source of funds rather than an objective for subsidization. Accordingly, debate now focuses upon how revenues can appropriately be raised to transfer resources within the sector in a way which compensates operators bearing a disproportionate burden of the USO costs. Essentially, the options differ in the breadth of the tax base proposed. Thus some argue that all telecommunications related activities, including regulated services, value added services and equipment should be subject to a tax, the proceeds of which go into a fund which is then distributed to firms incurring USO costs. Others propose that the tax base be confined to regulated services. Others still propose that net USO costs should be redistributed through surcharges on interconnect payments made by operators interconnecting with the operator discharging the universal service obligation. (This method is employed in Australia, and has some similarities with the system of access deficit contributions in use in the United Kingdom except that the latter is intended to redistribute losses associated with a sub-set of services rather than with a sub-set of customers.)

Standard arguments suggest that a broad tax base is likely to be desirable, because — other things equal — the lower associated tax rate will impose few economic distortions. There is also fairly general agreement that tying contributions to interconnection payments is potentially harmful as it will encourage inefficient bypass.

D. Price control

A final key aspect of telecommunications regulation concerns the procedures chosen to regulate prices, where such regulation is considered necessary because of continuing dominance of a wholesale or retail market by an incumbent. The major development here, pioneered in the United Kingdom, is the use of price caps. The key distinctions between price caps and traditional cost of service-making are two: first, price caps detach the firm's revenue stream at least from a period from its cost stream, thus creating a powerful incentive to reduce costs. Second, price caps typically permit the regulated firm greater flexibility in setting relative prices, subject to the overall price control. (For a useful survey, see OECD, 1995.)

These claimed advantages require, however, some qualification. In practice, price caps are set for a limited period, typically between three to five years. When the price control regime is renewed, one of the key factors determining the choice of the severity of the new regime is likely to be the current and expected level of profitability. This means that, compared with a price cap of infinite duration, the benefits accruing to a firm for cost reduction are substantially reduced — perhaps to as little as one quarter of the full amount, depending upon the period in the cycle when the cost reduction occurs and the policy adopted by the regulator. This observation is sometimes claimed to show that in practice price caps lack powerful incentive properties and are akin to rate of return regulation. Against this, it can be argued that a system in which the profit stream associated with a cost reduction attenuates gradually is much more akin to the operation of a competitive market than one in which it remains intact in perpetuity, and that the key distinction between classic cost plus rate of return regulation and a price cap is that in the former case the regulator is effectively endorsing *ex post* costs incurred by the company whereas in the latter the regulator has to make a forward looking projection of attainable productivity gains, which is then incorporated in the cap.

Standard price caps carry the risk that, if the potential productivity gain is mis-estimated, the company may make inadequate or excessive profits. In the latter case, public support for the system may be jeopardized. For this reason, several regulatory authorities, notably in the United States, have adopted an intermediate position known as profit sharing or sliding scale regulation in which returns above some specified level are shared between consumers and shareholders (Sappington and Weisman, 1995). A system of this kind has recently been endorsed by the Labour Party in the United Kingdom, although it is not certain whether its coverage would in fact extend to the telecommunications sector. Sliding scale regulation has clear disadvantages to be set against the greater fairness it is often perceived to exhibit. Any significant sharing of profits with consumers will reduce the returns to cost cutting efforts. Second, it may in practice be difficult to operate the systems symmetrically, so that firms are entitled to recover short-falls of profit from their customers through higher tariffs. As a consequence, the introduction of a system may not have the desired effect of reducing a firm's cost of capital. Finally, sliding scale regulation is inevitably more intrusive than a price cap, as the regulator has to be satisfied that the firm is not manipulating the system by making discretionary choices of when to recognize profits in its accounts.

As noted above, a second characteristic of price caps is that they typically give the firm greater flexibility to set relative prices. It has also been shown that, under certain circumstances, a firm subject to successive rounds of price cap regulation will eventually be drawn to charge Ramsey prices. (The conditions for this result are, however, fairly stringent — see Vogelsang and Finsinger, 1979.) However, regulators typically have an interest, often based on equity considerations, for restricting the price of access to the network. This implies a low, or even negative mark up on incremental cost of monthly or quarterly rentals whereas Ramsey pricing is normally taken to imply that the net cost to the network will be recovered predominantly through mark-ups on such charges. For this reason side constraints are often required, as well as an overall price cap and from the standpoint of pricing efficiency, the benefits of the price cap are diminished.

Although in practice most price caps have been applied to retail prices, attention is increasingly paid to their role in determining wholesale or interconnection prices. This immediately raises the issue of the articulation of wholesale and retail caps. The principal options are either separate wholesale and retail caps, with the latter constructed in a way designed to take account of the input price of network services; a primary reliance on a wholesale cap, with retail prices controlled mainly through competition; and a global price cap, recently proposed by Laffont and Tirole (1994). This would combine wholesale and retail prices in a single basket. Despite its closeness to the system with two separate caps, a global cap creates different, and more efficient, pricing incentives for regulated firms.

Laffont and Tirole recognize, however, that problems may arise under the global price cap if the dominant operator is able to carry out a price squeeze raising wholesale prices and reducing retail prices for certain competitive services. For this reason, the proposal for a global price cap is often accompanied by one to implement a competitive parity principle, similar to the ECPR described above, under which the retail prices charged by the dominant network operator must at least cover the wholesale price and the incremental cost to the retail services (Kahn and Taylor, 1993). In the United Kingdom, OFTEL is currently wrestling with the problem of co-ordinating wholesale and retail price caps. Practical experience of the problem will thus soon be available.

4. The regulation agenda

This section sets out a view of the agenda for regulators and students of the regulatory process. Many of the challenges facing both groups are related to problems of institutional design and of regulatory policy set out in the previous two sections. However, it is also useful, although dangerous, to speculate about how they will develop in the next few years.

At the institutional level, a key issue which most of the regulatory systems will face arises from the growing actual (rather than merely predicted) convergence of broadcasting, computing and telecommunications. Technical developments are constantly making available new methods of delivering broadcasting and telecommunications services in ways which expose the problems of combining a largely unregulated information technology sector with separately regulated broadcasting and telecommunication services. Thus consumers of entertainment services are indifferent to whether the delivery mechanism to their home is a cable television system, a conventional telecommunications network using ADSL for video on demand or analogue or digital terrestrial or satellite broadcasting systems. The development of telecommunications services offered over the Internet also blurs the distinction between computer and telecommunications networks.

These changes at least raise the issue of how to redesign the regulatory framework. To continue with separate regulation of essentially competitive services runs the risk of incoherence and arbitrage by firms across regulators. On the other hand, traditional broadcasting regulation has contained strong elements of control of content, which are largely absent from the armoury of the telecommunications regulator. One possible solution is to combine all economic regulation of the communications sector in a single agency, and transfer content issues to another body. This institutional arrangement would have the advantage of reflecting the likely different forms of political and social accountability appropriate to economic and content issues.

It is also worth mentioning that the expanding range of technologies capable of delivering particular services diminishes the market power of incumbent operators. Thus in principle it should bring forward the date when regulation of delivery mechanisms can be deregulated. However, other bottlenecks may emerge. Thus in Europe, considerable attention has recently focused upon potential monopolization of the conditional access system for pay television services. It is likely that in the future this will involve the use of a return path based upon a telecommunications network. Monopoly control of conditional access might thus be leveraged into the telecommunications market. This potential problem illustrates again the importance of regulatory institutions which take account of new technologies which cross traditional industry boundaries. A recent paper by OFTEL on convergence issues sets out how standard competition law and policy approaches can be applied to such issues (OFTEL, 1995a).

The points made above are a sub-set of a more general debate about the information super highway or in Europe's more humanistic language — the Information Society. It is satisfying to observe that early forecasts that expanding the bandwidth of communication channels between households and firms would lead to the emergence of a new natural monopoly have now largely been displaced. Instead, governments are increasingly recognizing the possibility that the "super highway" will in fact be a patchwork of alternative technologies, where the principal driving force is likely to be demand side factors, which will elicit competitive responses by firms. The debate has, however, focused attention again on the dangers of a division of society into information-rich and information-poor groups. However, recent proposals that public policy should focus on making broadband access available at local centres such as schools, health care organizations and libraries is an important potential method of alleviating the universal service problem at least in the short term.

References

Armstrong, M., C. Cowan and J. Vickers, 1994, *Regulatory Reform: Economic Analysis and British Experience* (MIT Press, Cambridge, MA).

Baumol, W. and G. Sidak, 1993, *Towards Competition in Local Telephony* (MIT Press, Cambridge, MA).

BTCE, 1989, *The Cost of Telecoms Community Service Obligations*. Report No. 64 (Canberra, Australia).

BTCE, 1995, *Telecommunications in Australia*. Report No. 87.

Cave, M., C. Milne and M. Scanlan, 1994, *Meeting Universal Service Obligations in a Competitive Telecommunications Sector* (DGIV European Commission, Brussels).

Cave, M., 1995, Development of the UK telecommunications industry through privatisation and deregulation, *Kaizi Bunseki* (*Economic Analysis*), Economic Planning Agency, Tokyo (forthcoming).

Goldberg, V., 1976, Regulation and administered contracts, *Bell Journal of Economics* 7:426-448.

Huber, P. *et al.*, 1993, *The Geodesic Network 2* (Geodesic Publishing Company).

Kahn, A., 1988, *The Economics of Regulation*. MIT Press.

Kahn, A. and W. Taylor, 1994, Comment on Baumol and Sidak, *Yale Journal on Regulation* 11:225-240.

Laffont, J.J. and J. Tirole, 1994, *Creating Competition Through Interconnection: Theory and Practice*, mimeo.

Levy, B. and P. Spiller, 1994, The institutional foundations of regulatory commitment: a comparative analysis of telecommunications regulation, *Journal of Law, Economics and Organisation* 10:201-246.

New Zealand Government, 1995, *The Regulation of Access to Vertically-Integrated Natural Monopolies* (Consultation Document, Ministry of Commerce and Treasury, New Zealand).

OECD, 1995, *Price Caps for Telecommunications: Policies and Experiences*, OFTEL.

OFTEL, 1995a, *Effective Competition: Framework for Action*.

OFTEL, 1995b, *The Costs, Benefits and Funding of Universal Service in the UK*.

OFTEL, 1995c, *Beyond the Telephone, the Television and the PC: Consultative Document*.

Sappinglon, D. and D. Weisman, 1995, *Designing Incentive Regulation for the Telecommunications Industry* (MIT Press).

WIK, 1994, *Network Interconnection in the Domain of ONP: Report to 4GXIII of the European Commission* (Bad Honnef, Germany).
Williamson, O., 1985. *The Economic Institutions of Capitalism* (Free Press).

CHAPTER 8

Comment: Marcellus S. Snow

Cave's outline of current issues and recent developments in telecommunications regulation is succinct, authoritative, and informative. Summarizing the policy challenges in a number of industrialized countries, he provides a particularly useful discussion of the present and recent past history of regulatory issues in the United Kingdom.

My intent is to complement Cave's contribution primarily by furnishing additional detail on developments and policy directions in the United States. In a number of instances I also wish to suggest alternative theoretical perspectives that are germane to the subject matter he has so ably presented and interpreted.

One cannot help but be struck by the pace and extent of telecommunications regulatory reform since the early 1980s in most if not all OECD countries, particularly in Australia, Japan, New Zealand, the United Kingdom, and the United States. Australia replaced a powerful government monopoly by a vigorous duopoly. Japan allowed pragmatism to prevail over long-established custom by privatizing NTT and opening it to competitors now numbering in the hundreds. New Zealand has gone further than any other country by not only privatizing and opening to competition its previously government-owned and bureaucracy-operated telecommunications system, but by seeking additional efficiency gains predicted by economic theory in conducting spectrum auctions and routinizing and simplifying market entry. The United Kingdom has pioneered by introducing price cap regulation; establishing OFTEL as a nationwide, specialized telecommunications regulatory; providing for duopoly as a matter of economic principle; and implementing telecommunications deregulation on the basis of carefully reasoned theory and policy arguments advanced by leading economists and other scholars.

The United States is often cited as an early "role model" of telecommunications deregulation. Indeed, impressive and long-lasting strides toward economic deregulation were realized as early as the 1970s in other sectors of the American economy (notably civil aviation and trucking) under the intellectual tutelage, patient advocacy, and administrative finesse of figures such as Alfred Kahn and adherents of the "political economy of deregulation" school [see Derthick and Quirk (1985)]. In addition, the divestiture of AT&T [see Snow (1995)] by the Modified Final Judgment agreed to 1982 — arguably the greatest rearrangement of private wealth ever undertaken by administrative fiat — was a precursor, if not an exact blueprint, of broad deregulatory actions taken by Japan and the United Kingdom in 1984 and 1985.

In many respects, however, the telecommunications landscape in America has changed less since the breakup of AT&T than it has in most of the other countries mentioned above, making the developments there even more impressive and sweeping by contrast. Here is why:

(1) AT&T and virtually all other telecommunications carriers in the United States had been privately held since the first commercial exploitation of telephony in the nineteenth century; thus, privatization was neither an additional issue nor another obstacle along the path of American deregulation.
(2) By virtue of privatization, a separation of operational and regulatory entities already existed in the United States.
(3) Labor union opposition to various deregulatory actions was less intense in the United States, where the unions involved were private-sector, relatively pragmatic bodies in an environment of sustained decline in union membership. By contrast, opposition to measures such as privatization and the introduction of competition in Australia, New Zealand, and the United Kingdom came from a much more powerful, unified, and ideologically homogeneous labour movement whose members were predominantly state employees.
(4) The option of "structural reform" as a weapon in the arsenal of monopoly policy — that is, the divestiture and reorganization of entire firms or their subsidiaries — was much more common and accepted in the United States than elsewhere. Indeed, the breakup of AT&T into the seven regional Bell operating companies differed more in degree than in kind from many previous examples, most of them traceable to initiatives of the Antitrust Division of the US Department of Justice.
(5) Finally, as Cave notes, the "multi-level" or "overlapping" regulatory model familiar in the United States for decades has evolved more recently and perhaps more tenuously in countries accustomed to centralized operation and regulation of telecommunications. Gerald Brock's (1994) views are relevant here. He holds that while the US system is apparently "chaotic", "disorganized", and shot through with "vague mandates and overlapping responsibilities", it is superior to those of other countries in its "ability to adapt to rapidly changing technological opportunities".

Indeed, much of the impetus for reform in such countries arises from a desire "to reform rigid and dysfunctional hierarchical structures".

The notion of natural monopoly is often used in a more precise and narrow sense in US policy discussions than elsewhere. The salient feature of natural monopoly, deriving from the fundamental insights of Baumol, Panzar,and Willig (1982) and others, is the fact that telecommunications cost functions are typically multi-product in nature. This richer and more complex use of the term allows subadditivity of the cost function to be the defining property of natural monopoly (among other things, making it amenable to econometric verification), and replaces the inherently single-output notion of average cost with the more nuanced concept of ray average cost.

Cave usefully interprets the regulatory authority as being ideally engaged in a sort of implicit or social contract, the dual purpose of which is "to protect consumers from excessive prices and to protect investors from expropriation of the value of their sunk investment". There are some very different classical views of the regulatory relationship, including the well-known arguments of Stigler (1971) and Peltzman (1976), who hold that poorly staffed and financed regulatory bodies are often "captured" by the very entities they were designed to regulate. Only time and experience will reveal if either of these somewhat polar views (idealistic and cynical) is useful in determining whether newly designed regulatory bodies succeed in promoting the often clashing interests of regulators, government, consumers, employees, entrepreneurs, investors, and taxpayers.

As Cave indicates, the four major regulatory issues affecting the telecommunications industry today are licensing (market entry); interconnection (access); universal service obligations; and pricing. One concept that has been useful in evaluating the issue of entry is the sustainability of natural monopoly. Contrary to single-output models, in a multi-product environment it is possible that financial incentives exist for profitable entry into markets that exhibit the characteristic property of natural monopoly, namely subadditivity of the cost function. The public policy dilemma raised by the possibility of such entry is whether an unsustainable natural monopoly — that is, one that is vulnerable to profitable entry by one or more potential competitors — should be artificially "sustained" by subsidies or entry proscription in order allow the least-cost technology to be used. The problem with such arguments is that they are static in nature and fail to capture dynamic considerations such as the emergence of new technologies and the improvement of existing ones.

Many of the issues highlighted by the introduction of duopoly in Australia (Optus against Telecom Australia) and the United Kingdom (Mercury against

British Telecom) have played out since the MFJ entered into force in the United States. Judge Greene, for example, has gradually relaxed "line of business" restrictions on the regional Bell operating companies. More to the point, Sprint and MCI — AT&T's major competitors in domestic long-distance markets — have effectively reduced AT&T's market share, while AT&T has complained loudly about the asymmetric regulation to which it has been subjected as the dominant firm.

In the context of licensing policy, Cave rightly favours placing the burden of proof on "those seeking to restrict entry". Competition is "much more likely to generate information about what customers want ..." This insight is reminiscent of Hayek's (1968) view of prices as an information system or "discovery procedure" that is at once efficient, decentralized, autonomous, and non-bureaucratic.

Finally, a discussion of universal service obligations concludes that "the case for [their extension] is at present weak". Yet Cave concludes his paper by asking whether differential access to the Internet or to the "information super highway" in general might bring about a "division of society into information-rich and information-poor groups". Certainly discussions of universal service obligations, given today's technologies, must touch on issues more subtle than who does not have a telephone at home?

References

Baumol, William J., John C. Panzar and Robert D. Willig, 1982, *Contestable Markets and the Theory of Industry Structure* (Harcourt Brace Jovanovich, New York).

Brock, Gerald W., 1994, *Telecommunication Policy for the Information Age* (Harvard University Press, Cambridge, MA).

Derthick, Martha and Paul J. Quirk, 1985, *The Politics of Deregulation* (Brookings, Washington, DC).

Hayek, Friedrich A., 1968, Der Wettbewerb als Entdeckungsverfahren, *Kieler Vorträge*, N.F. 56. Kiel.

Peltzman, Sam, 1976, Toward a more general theory of regulation, *Journal of Law and Economics* 19(2):211-240.

Snow, Marcellus S., 1995, The AT&T divestiture: a ten-year retrospective. In: Donald M. Lamberton, ed., *Beyond Competition: The Future of Telecommunications* (North-Holland, Amsterdam) 207-226.

Stigler, George J., 1971, The theory of economic regulation, *Bell Journal of Economics and Management Science* 2(1):3-21.

CHAPTER 8

Comment: David Cracknell

The contribution of the regulator has been and will continue to be central to the way in which telecommunications pricing develops in the UK. In a 1994 consultative document, Oftel, the UK regulator stated that his objectives were:

- To progress towards the goal of efficient, free and competitive multi-operator markets with minimal regulation and market determined prices;
- To cause prices to evolve in a timely way to meet the needs of multimedia service markets;
- To ensure a fair distribution of costs and benefits among both personal and business customers.

This statement shows that the scope of regulation has advanced considerably in 10 years. Littlechild in his report on regulation of the privatized BT, stated that "Competition is indisputably the most effective means — perhaps ultimately the only effective means — of protecting consumers against monopoly power. Regulation is essentially a means of preventing the worst excesses of monopoly, it is not a substitute for competition. It is a means of "holding the fort" until competition arrives." As we have seen, the UK is now widely recognized as being the most open telecommunications market in the world, as stated by the US congressional committee on technology assessment in 1992.

One might, therefore, conclude that the regulatory regime had met the needs placed upon it in 1984. Rather than decline as competition has developed, however, the regime has extended its scope by:

- Steadily tightening the price cap. This started out at 3 percentage points below inflation in 1984, and now runs at 7.5 points below inflation until 1997.
- Imposing additional regulatory controls, such as the cap on the bill of the median personal customer in 1989, and the inflation cap on all prices (except access rentals) in 1993;
- Bringing international call prices under regulatory control in 1991. As we have noted, BT now has only a 74% share of this market;
- Introducing price floors which "institutionalize" the cross subsidy between access rentals and call prices;

- Ensuring that asymmetry persists in the market. BT's cable TV competitors are allowed to provide both entertainment and telephony services, while BT remains banned from the broadcast entertainment market.

This assortment of measures shows the differing, and often conflicting stakeholders in regulatory policy. They might be characterized as:

- Consumers. The movement of the telephony bill of an average personal and business customer at 1995 prices (based on current usage patterns) over the last 25 years shows that the regulation has ensured a smooth downward pattern, as distinct from the rather erratic movement in the previous 15 years.
- Competitors. As we have noted, the regulatory regime has encouraged around 150 operators in the UK market. Some have been assisted more than others — cable TV companies undoubtedly benefit from the restrictions preventing BT from competing in the broadcast entertainment market. MCL, BT's first competitors, have, however, recently announced a degree of retrenchment in the personal and small business market, which has been attributed by them, and other analysts to the severity of the current BT pricing regime.
- Social needs. Through its restrictions on rebalancing, and increasingly on particular prices, the regime obliges BT to cross subsidize low users, at the expense of high users. Although this also benefits competitors by providing opportunities to win market share by "cherry picking" very profitable customers it presumably also reflects social policy and the need to provide subsidized telephone service to the lowest users.

It also highlights some other anomalies in the UK regulatory regime:

- As head of a non-ministerial government department, to whom is the regulator answerable? How can democratic control be exerted over the regime?
- What are the regulator's objectives? How are the needs of consumers and competitors balanced?
- What are the boundaries of the regulator's discretion?

In a recent speech to an FT conference, BT's chairman provided a vision of how the regulatory regime might move forward to clarify some of these issues. He suggested that it should:

- Commit to deregulate all retail activity from 1997. This would involve no price caps, and no constraints on pricing structures beyond normal competition law;
- Impose specific time limits on market entry supports for new competitors;
- Provide a stable and explicit regime for interconnection services, that applies to all licensed operators.

Introducing this degree of clarity into regulatory objectives, would make the role and accountability of the regulator much clearer than at present, to the benefit of all stakeholders of the regulatory regime.

CHAPTER 9

Telecommunications regulation: an Australian perspective

A.J. Shaw[1]

Martin Cave's chapter highlights areas that have concerned Governments undertaking telecommunications reform, with a focus largely on developments in the United Kingdom and Europe. This contribution provides a comment from an Australian perspective.

Governments around the world largely face the same types of problems in undertaking telecommunications reform, and each can learn a great deal by drawing on the experience of other countries. Nevertheless, national conditions and concerns will influence the ultimate goals and the best path to achieve them. Australia has consciously drawn on overseas experiences in developing its own particular approach to the reform process.

My comments focus on six key regulatory issues which have dominated the debate on Australian telecommunications reform since 1987: institutional arrangements; carrier licensing; interconnection and access; anti-competitive conduct; universal service; and price control. Starting with the first reforms announced in 1988 and covering the 1990-91 and proposed post-1997 arrangements, the paper shows the changing nature of the Government's response to each issue over time as the reform process has progressed. A brief description of each reform package is provided in Box 1.

Box 1: Overview of Communications Reform in Australia

1988	May: Evans Statement — clarification of carrier rights; carrier corporatization; capping of Telecom prices; establishment of AUSTEL; competition in value-added services and customer equipment; reviews of Telecom's CSOs and existing ownership and structural arrangements.
1989	July: *Telecommunications Act 1989* and *Australian Telecommunications Corporation Act 1989* commence giving legislative effect to 1988 statement.

1. The assistance of Philip Mason in preparing this paper is gratefully acknowledged. The paper was prepared in January 1996 and reflects the status of Government policy at that time. Some of the policy principles for post 1997 are expected to be changed by the new Government elected on 2 March 1996.

1990	November: Beazley Statement — merger of Telecom/OTC; sale of AUSSAT as basis for second network and mobile operator; third mobile operator; full resale; extensive pro-competitive safeguards including interconnection, access and preselection; enhancement of AUSTEL; equitable and transparent industry funding of USO; continuation of price caps and other consumer safeguards; 1997 sunset on market entry restrictions.
1991	July: *Telecommunications Act 1991* commences.
	November: Optus selected as second carrier.
1992	June: Optus begins offering mobile services.
	September: Spectrum management reforms announced — establishment of Spectrum Management Agency (SMA); changes to spectrum administrative arrangements; introduction of scope for market-based spectrum management, including trading, in selected bands.
	October: *Broadcasting Services Act 1992* proclaimed — new service categories reflecting services' ability to influence public opinion; new approach to broadcasting planning; market-based allocation of commercial licences; new concept of "control"; introduction of subscription broadcasting; industry-based programming standards; establishment of Australian Broadcasting Authority (ABA).
	November: Optus begins offering long distance and international services.
	December: Vodafone licensed as third mobile carrier.
1993	July: *Radiocommunications Act 1992* commences.
	September: Vodafone commences operations.
1994	March: Satellite pay TV licensing process completed.
	May: Terms of Reference for Post-1997 Review released.
1995	July: *Telecommunications (Carrier Associates) Direction* provides for regulation of cable pay TV operators, including provision of access and interconnection.
	August: Lee Statement on post-1997 arrangements — open market entry; removal of many carrier privileges; industry-based access undertakings regime overseen by Australian Competition and Consumer Commission (ACCC); anti-competitive conduct flexibly regulated by ACCC; merger of AUSTEL and SMA; introduction of industry codes of practice; improvement of USO arrangements; continued price caps and other consumer safeguards.
1996	Introduction of *Telecommunications Bill 1996* (anticipated)
1997	July: *Telecommunications Act 1996* and open competition commence (anticipated)

During this period Australia has also completed major reforms of its broadcasting and radiocommunications legislation. The result has been a massive change in Australian communications over the past eight years.

Institutional arrangements

While the institutional arrangements for the administration of regulation are of secondary importance to the actual regulatory provisions themselves, they nevertheless influence the way those arrangements work and it is sensible to discuss them at the outset. Australia's experience in this area has been that institutional arrangements need to change with the environment and the regulatory framework.

Prior to 1989, the publicly owned domestic and international monopoly carriers (now combined to form Telstra) exercised significant regulatory powers. With the decision in 1988 to introduction competition in value-added services and customer premises equipment and clarify the domestic and international carriers' respective monopolies, it was clear that it would be inappropriate for the publicly owned incumbents to continue to exercise those powers and that an independent regulator would be required. To maximize independence, the Australian Telecommunications Authority (AUSTEL) was established as a statutory authority. AUSTEL's original functions were to protect the reserved rights of the carriers and promote carrier efficiency, protect competitors and consumers from unfair carrier practices and undertake technical regulation. While the concept for AUSTEL came from the United Kingdom, the powers conferred on it differed significantly from those of OFTEL. Under the Australian model, the Minister held the main policy levers, with AUSTEL being the administrator of that policy. For example, while OFTEL sets price caps, AUSTEL simply administers caps set by the Government.

Significant changes to AUSTEL were part of the 1990 decisions to merge Telecom and OTC and to introduce a duopoly general carrier arrangement. Carrier competition created a much greater role for AUSTEL, for example in relation to interconnection, dominant carrier pricing, accounting separation, consumer protection and administration of the universal service arrangements. Most importantly, AUSTEL's power to arbitrate access agreements between carriers where the carriers were unable to agree became fundamentally important. Carriers were given no right of appeal on the merits of AUSTEL decisions. AUSTEL responsibility for numbering and administration of the universal service levy enhanced carrier equity, while its reporting on competitive safeguards and carrier performance improved Government scrutiny of the reform process. Significantly, AUSTEL was also given the more active role of promoting competition, in contrast to that of simply protecting competitors from unfair practices and promoting fair and efficient market conduct as had been the case under the previous 1989 regime.

In August 1995 the Government reaffirmed its policy of opening up the market to further competition from July 1997. In doing so it indicated that its policy was for post-1997 legislation to facilitate a smooth transition to an eventual reliance on general competition policy law by aligning telecommunications-specific law with trade practices law to the maximum extent practicable. In support of this, responsibility for administering telecommunications-specific competition policy is to be transferred from AUSTEL to the general competition regulator, the Australian Competition and Consumer Commission (ACCC). This change is consistent with the ACCC's general responsibility for competition and access regulation and reflects the view that telecommunications-specific regulation relating to competition matters is best administered by the national competition regulator so that the broad approach to competition regulation across all Australian industries is consistent. This arrangement also avoids carriers facing the "double jeopardy" of having competition matters regulated by both an industry regulator and a general competition regulator. Ultimately, the regulation administered by the ACCC is enforceable through the Federal Court.

After 1997, the remaining elements of AUSTEL (principally dealing with consumer safeguards, licensing and technical regulation) will be merged with the SMA to form the Australian Telecommunications and Spectrum Management Authority (still to be known as AUSTEL). This will streamline regulation administration and allow convergence issues to be better addressed. The "new" AUSTEL will be responsible for carrier licensing, consumer protection, spectrum management, technical regulation, numbering and management of the universal service arrangements. AUSTEL's telecommunications-specific consumer protection functions will supplement those of the ACCC and the industry-based Telecommunications Industry Ombudsman (TIO).

In developing the post-1997 institutional arrangements the Government did not accept arguments that convergence in electronic communications required a cross-sectorial regulator dealing with communications infrastructure, carriage and content. Rather, the Government took the view that content raises such distinct issues that its regulation should continue to be administered by a separate body, the Australian Broadcasting Authority.

An important aspect of the post-1997 institutional arrangements, as indeed with the broad approach to post-1997 regulation, is the preference to rely, where practicable, on general, economy-wide law with supplementary telecommunications-specific arrangements only adopted to meet special circumstances.

This Australian trend goes against the trend identified by Cave for industry-specific regulators to discharge the main regulatory functions but increasingly view their task as one of implementing competition law and policy. That assessment would accurately describe the current Australian regulatory situation but this is to change with the move from the duopoly to open competition. While these changed institutional arrangements reflect the reducing need to have an industry-specific regulator with a pro-competitive charter as competition develops, as noted above this approach was also influenced greatly by a desire to more closely align telecommunications access policies with those contained in the new *Competition Reform Act 1995*.

Carrier licensing

Australia, like many other countries, has followed a progressive approach in introducing competition into all telecommunications activities.

The process, depicted in Box 1, started with the introduction of competition in customer premise equipment and value-added services in 1989. This was followed by the licensing of a second general carrier and mobile carrier (Optus) in 1991 and a third mobile carrier (Vodafone) in 1992, but with a commitment to move to more open competition on 1 July 1997. Full resale was also introduced in 1991.

The decision in 1990 to adopt a transitional general carrier duopoly reflected both past circumstances and future goals. In 1990 the Government accepted as a fundamental tenet that competition in telecommunications was the best way to encourage efficiency, innovation and consumer responsiveness and in the long term open competition was most desirable. A key consideration was the view that it was necessary to nurture a strong competitor to the incumbent monopoly carriers and that this could best be done by limiting additional entry for a specified period. As Cave notes "one way of visualizing the process of entry into a market which was previously an exclusive one is as a race between the incumbent's capacity to adapt itself to the new market conditions ... and the entrant's ability to accumulate the necessary assets to survive in a market which by definition is new to it". This was clearly recognized in Australia and, with a view to serving Australia's long term interests, the Government provided the industry with a period of adaptation.

Providing the transitional period has enabled Telstra to restructure itself to meet the demands of the competitive market place. This was essential given the Government's multiple goals of retaining Telstra in public ownership and making it a flagship of international competitiveness and Australian investment offshore. Similarly, the transitional period enabled the new operators, Optus and Vodafone, to concentrate on competing with Telstra and each other and establishing their market presence.

The revenue collected through the sale of carrier licences in 1991 and 1992 reflected the enhanced opportunities because of the limit on market entry, but this was a fortuitous consequence rather than a prime motivation. The main focus of the Government was on the international market place and on the domestic competition policies required to meet its national goals. This can be contrasted with the approaches taken in the United Kingdom and New Zealand.

Unlike Britain's decision to introduce a duopoly, Australia's decision to limit entry had a time limit from the very start. While the new carriers were given a period to establish themselves, they were given a strong incentive to do so rapidly and effectively and not settle into a "cosy duopoly".

After 1997, the Government's policy is to adopt a different concept of "carrier" from that now in place. There will be no requirement that persons installing telecommunications infrastructure seek licensing as a carrier. Rather than being a person licensed with special rights, a carrier will be a person who, by virtue of their control of access to "significant" telecommunications networks or components of networks which are used to provide services to the public, will have special obligations. This will assist in "regulating" convergence because all types of infrastructure used in telecommunications networks or components of such networks (cable, satellite, radio — be they fixed or mobile) will be covered where they are used to provide service to the public. The Government has decided there will be no numerical restriction on the number of carrier licences issued, and no entry hurdles in relation to financial and technical capabilities. Licences will be issued for an administrative-based fee.

This post-1997 approach reflects the view that investors are capable of making their own market entry decisions and that an open market removes any justification to collect licence fees from carriers. It also reflects a view that there is no need for continued special measures to promote entry — a position that accords with the views expressed by Cave.

Interconnection and access

The importance of interconnection and access in telecommunications in Australia was recognized well before the introduction of network competition. While the *Telecommunications Act 1989* clarified the respective monopolies of Telecom, OTC and AUSSAT (the satellite operator), it also gave the carriers the right to request AUSTEL to direct another carrier to provide access to a network or facility for its use for an amount agreed or otherwise determined by the industry regulator. Connection rights were also given to value-added service providers.

The decision to allow new entrants in 1990 was accompanied by a decision to create a more robust and effective interconnection and access arrangement. The Australian arrangements, contained in the 1991 Act, were developed drawing on overseas experience, with the approach finally adopted combining the best features of a number of regimes — but with many features drawn from North American approaches. The New Zealand scheme of relying on commercial negotiations was rejected as too risky: the UK approach was seen as having contributed to the relatively slow growth of Mercury.

In addition to clarifying and expanding the rights of interconnection and access in the 1989 Act, the *Telecommunications Act 1991* enabled the Minister to determine principles and actual charges to apply in interconnection and access. (General licence conditions provided more detailed supplementary access rights.) The Government used the power to require Telecom/OTC to provide interconnection at directly attributable incremental cost and prescribed certain initial interconnection charges. This was designed to set charges at a level the new entrant would be expected to pay if it possessed an established market presence, with a view to promoting growth in the new carriers' market shares. Once initial charges ceased, access charges were negotiated on a commercial basis. The principles determined by the Government still provide the basis for carriers' access agreements and for AUSTEL arbitrations, and will continue to do so until new access arrangements come into force in July 1997. Access agreements are assessed by AUSTEL against the determination and registered, thereby exempting them from general trade practices law. These arrangements are important also in differentiating carriers from service providers.

Cave has indicated his view that symmetry in the regulation of interconnection is important and refers, among other things, to non-discrimination. While a dominant carrier is generally prohibited under current Australian legislation from discriminating at the retail level to prevent anti-competitive conduct and when supplying basic carriage services to produce higher level services, there are no statutory non-discrimination rules applying to carrier-carrier transactions. Rather, under the overall access arrangements, requesting carriers have the right to obtain access at a fair and reasonable price, if not an identical price to the price the supplying carrier pays to itself. This recognizes the supplying carrier's property rights. As discussed below, there will be some changes to these provisions after July 1997.

Unlike some other countries, Australia also mandated the introduction of preselection, requiring Telecom/OTC to upgrade its exchange capability as necessary. Such a requirement provides obvious advantages for a new entrant in marketing itself to the customer base and its availability was considered a fundamental element of an equitable regulatory regime. Preselection was also justified as a consumer safeguard. Growth in Optus' share of long distance and international business was also helped by an AUSTEL decision to hold ballots on a region by region basis. In the ballots, existing Telecom subscribers could choose to remain with Telecom or become Optus customers. Under the ballots between 4% (in Bendigo) and 18% (in Sydney) of customers have preselected Optus, giving it a national average of around 14% of customers balloted (Budde 1995:54).

Under the 1991 Act, full resale was allowed and service providers were given rights to connect with carriers' networks. Unlike carriers, however, their level of connection and the price was a commercial decision for carriers, subject to the disciplines of the commercial market place. (There is no right of appeal but it is proposed that this will change post-1997.)

The comprehensive interconnection and access arrangements and other competitive safeguards put in place in 1991 have made a significant contribution to the rapid growth of competition. By 30 June 1995, four years after the legal commencement of competition, Optus had secured an estimated 14.6%, 22.8% and 34% of national long distance, international and mobile revenues respectively, while Vodafone, which only legally commenced operation in July 1992 and only supplies digital mobile services, had secured an estimated share of 6.5% of mobile revenues (Budde, 1995:61, 100). Non-carrier service providers have captured around $1,268m (about 8.5%) of the total $15,018m Australian market (Budde 1995:61). Such figures contrast favourably with the growth of new carriers in Britain and New Zealand.

The interconnection and access arrangements to be implemented after 1997 will be different in appearance to those put in place in 1991. These changes reflect four main influences: the need to effectively facilitate interconnection and access among a larger number of carriers; the associated objective of giving the industry itself greater responsibility for managing interconnection and access; diminution of regulatory differences between carriers and service providers; and the desire to better align telecommunications access arrangements with general access arrangements included in Australian competition law by the *Competition Policy Reform Act 1995*.

Under the proposed arrangements, any-to-any connectivity and open access for service providers will be guaranteed by a requirement that all carriers participate in a Telecommunications Access Forum which will develop an Access Code. Against this Code, carriers will then give access undertakings which will be approved by and registered with the ACCC. Each undertaking would set out the terms and conditions on which a carrier is prepared to offer interconnection and access to carriage services, networks and components of networks. As such, it will provide the basis upon which service providers, including other carriers, will negotiate their access arrangements. Where parties cannot conclude an agreement they will be able to seek arbitration by the ACCC.

The most significant difference between the 1991 and 1997 approaches is the shift towards the access arrangements now in the general competition law, although there will still be many specific legislative provisions that relate only to the telecommunications industry. The proposed arrangements will require that full consideration is given to the legitimate business interests of the carrier in any dispute resolution process — which is vital in promoting sustainable competition over the long term. At the same time, however, the arrangements recognize the public interest aspects of an open access policy. In this context, an important element of the arrangements is the absence of any legal distinction between interconnection and access between carriers and carriers and service providers.

These new arrangements will satisfy most of the criteria identified by Cave as important in this area: guaranteed any-to-any connectivity and a mechanism for the regulator to require equality of interconnection services, unbundling of interconnection services, and to intervene on the pricing of those services if commercial negotiation fails.

Regulation of anti-competitive conduct

An important regulatory issue for Australia, but one that Cave does not comment on extensively, is the regulation of anti-competitive behaviour, particularly by a dominant carrier. In Australia, this by definition has until recently meant the incumbent carrier, Telstra. The particular importance of this issue in Australia reflects the nature of Australian competition law with its court-based processes which cannot always be relied on to provide timely remedies to anti-competitive behaviour.

In introducing competition in 1991, Australia decided that the general provisions of trade practices law, with its emphasis on anti-competitive intent or purpose and court based processes, would not be sufficient for the newly competitive telecommunications industry, characterized by a powerful, well established incumbent and new players. Accordingly, specific rules, considered more appropriate to the telec-ommunications market and more conducive to the introduction and promotion of competition, were adopted to supplement general trade practices law. Most importantly, the dominant carrier in a market (in practice, the incumbent, Telstra), was subject to special pricing and tariffing requirements. On the basis that discrimination and off-tariff charging by a dominant carrier would in themselves be evidence of anti-competitive behaviour, the dominant carrier was prohibited outright from discriminating between acquirers of like services and charging off-tariff except in special circumstances. These rules were underpinned by tariff filing and accounting separation, with AUSTEL oversight.

By 1993 it had become apparent that the dominant carrier pricing rules were not working as fully intended and they were amended in 1994 to give the dominant carrier in a market greater pricing flexibility while improving AUSTEL's ability to deal with anti-competitive pricing. First, the amendments expanded the exemptions from the anti-discrimination provisions, for example, allowing for tariffs of broad appeal. Second, as a complement to the expansion of exemptions, the amendments empowered AUSTEL to disallow a dominant carrier tariff if it was anti-competitive. The practical effect of the changes was to allow a dominant carrier to file certain tariffs which might otherwise be considered discriminatory but subject them to a further anti-competitive effects test. These changes show an important development in the regulation of anti-competitive conduct in telecommunications, reflecting a shift away from strict prohibition of discriminatory pricing and the presumption of anti-competitive behaviour towards regulation of anti-competitive effect itself.

In moving to 1997, the Government wanted to implement a regime that was less onerous and more consistent with general trade practices law, but still dealt effectively with possible abuse by a dominant carrier of its market power. (Importantly, in 1995 the prohibition on price discrimination under general trade practices law was repealed in implementing the national competition policy.) As competition would still be in its comparatively early stages, speed of regulatory action remained critical. There was also a concern that carriers with market power should not be able to use time-consuming processes, including litigation, to frustrate the entry of new players. At the same time, those players should not be overly restricted in responding to new commercial pressures.

In this context, unlike the 1991 Act, post-1997 regulation of anti-competitive conduct in telecommunications will not involve explicit prohibitions on particular conduct. Rather, after 1997 the ACCC will have the power to direct a telecommunications operator with a substantial degree of power in a telecommunications market to cease or not to engage in conduct where they take advantage of their market power for the purpose, or with the likely effect, of substantially lessening competition, or potential competition, in that or in any other telecommunications market. This direction power is a telecommunications-specific supplementation of the general remedies otherwise available to the ACCC. The ACCC will also have the ability to exempt specified conduct from the scope of the telecommunications-specific, but not general trade practices, anti-competitive conduct rules. Where directed to do so by the ACCC, a telecommunications operator with a substantial degree of market power will have to file the tariffs it is charging. This will ensure tariff information is available where there are concerns about anti-competitive behaviour or it is otherwise in the public interest. These ACCC powers will be supported by information gathering and accounting separation powers.

The Australian trend, therefore, has been to adopt strong anti-competitive behaviour provisions with the initial introduction of competition but move to provide carriers in a position of substantial market power with increased flexibility as the market becomes more competitive. At the same time, Australia will maintain strong provisions to allow the regulator to deal rapidly with conduct regarded as anti-competitive should it arise. The net result is likely to be an increase in the complexity of legislative provisions and more discretion on the part of the regulator.

Universal service

The Australian experience indicates that the introduction of competition is much smoother when accompanied by policies that safeguard universal service objectives.

In considering options for telecommunications regulation in 1987 and 1988 it became a matter of concern that, although universal service was essentially a Government mandated social equity objective, the level of resources involved were unknown and were outside the Government's direct control and accountability provisions. This blind spot had implications for the operation of the industry as a whole and for Telecom's operation as a corporatized entity. In response it was decided that the Bureau of Transport and Communications Economics (BTCE) would examine and report on the costs and cross-subsidies associated with meeting identifiable community service obligations.

The BTCE, adopting an "avoidable cost-revenue forgone" methodology, estimated the maximum annual cost at the time to Telecom of providing universal service (and some other minor community service obligations) to be \$270m (BTCE, 1989:xvii). This compared with a maximum estimate by Telecom, prepared using a fully distributed cost methodology, of \$800m. By establishing a costing methodology and the cost of universal service, the BTCE helped resolve universal service related arguments against the introduction of competition and laid down a basis for dealing with the costs of universal service in more open market arrangements.

In moving to competition, it was recognized early that for Telstra not to be unduly disadvantaged the burden of providing universal service should be shared with its competitors. This concern was complemented by growing public awareness of the importance of promoting and safeguarding high levels of consumer access to telecommunications. As in other countries where the provision of universal service has depended on cross-subsidies within monopoly providers, it was recognized that such an arrangement, by placing an inequitable burden on the incumbent carrier, could not be sustained in a competitive regime. Accordingly, the 1991 Act defined the universal service obligation and provided for industry funding of losses incurred in its provision, with carriers contributing in proportion to their share of total timed traffic. Basically the obligation is to make the public switched telephone service — including a non-switching telephone and payphones — reasonably accessible to all people in Australia. The service to which the obligation relates may be changed by regulation. For maximum flexibility, the Act enables any carrier to be the universal service carrier but because of its market presence, Telstra was declared the universal service carrier.

The universal service carrier is required to identify areas in which it makes a loss supplying services under the universal service obligation. Losses are calculated according to the "avoidable cost-revenue forgone" methodology developed by the BTCE, with the universal service carrier's calculation being subject to scrutiny by AUSTEL and contributing carriers. Losses are funded by all carriers with payments being channelled from debtor carriers to creditor carriers (that is, universal service providers) via a Government administered Universal Service Fund.

Together these arrangements promote three important objectives: the economical and efficient delivery of the services, transparent costing and equitable funding.

While some aspects of the costing methodology are currently being refined through a process being undertaken by AUSTEL and the carriers, the existing arrangements are to continue after 1997 with only slight changes. These changes are designed to simplify or otherwise improve administration of the arrangements and better align it with the post-1997 open Australian market. For example, after 1997 universal service carriers will be required to prepare plans about fulfilling their obligations, prices for services supplied by all carriers providing service under the universal service obligation will be able to be controlled and funding will be based on carrier revenue from carriage services, rather than timed traffic.

Australia has been well aware of the arguments highlighted by Cave about not tying contributions for universal service to interconnection payments, which is why a broader tax base was adopted from the outset. The move from timed traffic to carriage revenue recognized the increasing difficulties with a timed traffic concept in the broadband environment while acknowledging that the tax base should be as broad as possible.

Price control

Price capping of Telstra prices, complemented by notification and disallowance provisions, was introduced in the 1989 reforms. Price caps are determined by the Minister of State with portfolio responsibility and compliance with the caps is monitored and enforced by AUSTEL. Prior to 1989, Telecom and OTC prices for standard or basic services were subject to Ministerial approval following consideration by the then Prices Surveillance Authority (PSA).

Price control was adopted to provide downward pressure on prices in the absence of competition, providing both an incentive for efficiency improvements and a means of ensuring consumers benefit from efficiency gains. Price capping — requiring prices to fall by a CPI-X factor — was adopted because of the incentives it gives a carrier to improve its efficiency. Notification and disallowance, a modification of the 1989 approach, was adopted to deal with charges for certain sensitive, socially important services like payphones and directory assistance. Australia consciously considered and rejected the use of rate of return regulation as prevailed in the United States because, while preventing excessive profits, it provided little incentive for improving efficiency.

The price control arrangements introduced in Australia in 1989 have been retained under both the 1990-91 and 1995-97 reforms, even though those reforms have included the progressive introduction of competition. The retention of price control reflects the Government's ongoing concern to ensure that, while competition is developing, there continues to be downward pressure on prices to the benefit of all consumers and overall industry efficiency, particularly in markets where competition may be slow to develop, such as the local access and call markets. The application of price caps on Telstra, as the price setter, has a flow-on effect on the whole industry. In moving to an open regime after 1997, the Government has decided that price caps determined to apply from 1 January 1996 will continue until 31 December 1998, 18 months into the open competition period. Price controls for 1999 and beyond will be decided in 1997-98 with due regard to prevailing market conditions, including the progress of competition.

Both the 1991 and proposed post-1997 legislation require any operator providing local call services to provide residential customers with the option of untimed local calls. This reflects the importance that the community and the Government attaches to this policy. However, neither the 1991 nor proposed post-1997 legislation has any requirement for uniform (that is, geographically averaged) pricing. The overall emphasis of pricing regulation has been to ensure overall price affordability, rather than promote superficial equity through uniform pricing requirements. The better approach is to address any undue social disadvantages from non-uniform pricing through the price caps, which can (and do) include rebalancing constraints, and through direct assistance.

The future regulatory agenda

Predicting future regulatory issues is notoriously difficult, but it is expected that the issues likely to be on Australia's regulatory agenda are those identified by Cave. As Australia works towards the 1 July 1997 starting date for open competition against the background of the Government's policy principles, it is likely that many issues will relate to implementation and administration rather than underlying policy. Some of the issues likely to be on Australia's regulatory agenda over the next few years are as follows.

Interconnection and access will remain contentious areas because they are of such fundamental importance to the operation of a competitive telecommunications industry and there must be no chance of the arrangements failing. Carriers and service providers will inevitably test the new arrangements and, like AUSTEL in the past, the ACCC may be required to make some fine judgements.

Dealing with anti-competitive behaviour is also likely to be contentious, particularly in settling the precise form of the ACCC's directions powers and with industry participants testing the limits of the new mechanisms.

With market entry thrown open and wireless service possessing comparatively low barriers to entry, it is easy to envisage *access to spectrum* becoming an increasingly important issue. Again, it will be a potential area for conflict between incumbents, eager to keep and augment existing spectrum holdings to accommodate growth and exclude new entrants, and new entrants, eager to get a piece of the action.

As in most advanced economies, *consumer access to telecommunications and pricing* have become, and are likely to remain, major public issues. The wish to ensure competition enhances access rather than jeopardizes it, the roll-out of new infrastructures (in Australia roll-out of hybrid fibre-coaxial networks only commenced in 1995) and growing familiarity with higher capacity uses, especially the Internet, will create added pressure in this area. The access issue is likely to continue to be seen as involving the physical availability of infrastructure and the affordability of the services. Increasing use of mobile services may focus attention on their coverage, accessibility and pricing.

Access to content, as well as the services to access it, is being debated. An important and recurrent issue for Government will be determining the appropriate level of access it needs to ensure.

Content regulation is a perennial issue in communications policy and is likely to remain so as long as people's tastes differ. With competition and Government intervention providing increasing access to on-line services, including the Internet, some community concern about the content being accessed is to be expected. As usual, Government will be expected to adjudicate between competing community expectations in this area.

Like content, *privacy* is another perennial of telecommunications regulation, probably because it is so interlinked with commercial and technological developments. As commerce and technology change, new privacy concerns emerge. It is becoming increasingly recognized that privacy issues associated with telecommunications (like the collection of call information and telemarketing) are not necessarily telecommunications specific but manifestations of general privacy issues in the telecommunications context. The challenge here is to develop a robust, adaptive privacy framework capable of dealing with and accommodating change in telecommunications and other industries.

The conclusion of the General Agreement on Trade in Services and the APEC Bogor Declaration will have important implications for Australian telecommunications policy as it relates to *international matters*. Increasingly, these agreements will throw up issues for policy makers and provide a framework within which other issues will need to be addressed.

Conclusion

With new problems, issues and circumstances continually emerging, a country's policy and regulatory framework must be able to change as needs dictate. Since Australia commenced its program of communications reform in 1988 it has made significant progress. The extent of that progress in key areas will be evident from the preceding discussion. Important among the factors contributing to Australia's success in this area over the past eight years have been broad consultation in the development of policy; the formulation of sound underlying principles, giving a high degree of policy predicability; appropriate regulatory flexibility and reasonable regulatory discretion within the overall policy framework; and ultimately, a commitment to change when the need for change became clear.

References

Budde, P., 1995, *Telecommunications Strategies 1995/96* (Paul Budde Communication, Bucketty Way).

Beazley, K., 1990, *Micro-Economic Reform: Progress Telecommunications* (Department of Transport and Communications, Canberra).

Bureau of Transport and Communications Economics, 1989, *The Cost of Telecom's Community Services Obligations* (Report 64) (AGPS, Canberra).

Department of Transport and Communications, 1992, *Radiocommunications Spectrum Management Reform* (Department of Transport and Communications, Canberra).

Department of Transport and Communications, 1993 *Broadcasting Reform: A New Approach to Regulation* (Department of Transport and Communications. Canberra).

Evans, G., 1988, *Australian Telecommunications Services: A New Framework* (AGPS, Canberra).

Lee, M., 1995 *A New Era in Australia's Telecommunications* (Department of Communications and the Arts, Canberra).

CHAPTER 10

Social dimensions

David Allen

In the workshop from which this volume grows, views continued to divide into what seemed two poles. The microcosm of the workshop, it appeared, brought to some focus a normally less-choate tension in the debate on communications and information. Economics has enjoyed some success, and even a dominant place, in both the intellectual and the policy dialog. But when views from an array of other disciplines (and other policy persuasions) enter the discussion, the tension begins to take form.

The subject at the crux between these two sets of views appears to be universal service, at least that is the subject to which attention repeatedly and finally turns as the two sides discuss. A supporting leitmotif is intellectual property.

Universal service as the dividing, and uniting, subject points, I believe, to the fundamental which separates the two sets of views. Though economics is most of all a study of aggregates, it is founded on an atomistic construction of those aggregates. The individual, and individual choice — even if choice by an individual family or firm — are the stuff from which economics thinking is built.[1] Views from "the other side" will, on occasion and for a large variety of reasons, be concerned with some social group as a starting point for analysis. Whether that is a nation in political science, a social group *per se* in sociology, an ethnic minority in anthropology, a work team in management science, a community in philosophy, or in geography, or an era in history, the unit is a social aggregate.[2]

Much mischief may devolve when two (colliding) frames of thought differ by this fundamental. Universal service draws the contrast sharply into focus. The market in economics is ordinarily considered in terms of penetration *up to* some level. From the other approach, concern for the welfare of a group may ask whether all members are accounted — the starting point is the group as a whole, and counts *down from* there. Universal service for a communications network precisely marks the point of divide between the two styles of thinking.

(Of course there are not just two poles of thought — thought in economics is comprised of many competing voices and there are very many other voices on "the other side". It is only when two larger frames of thought begin to bump up against each other that contrasting poles appear to take shape.)

Intellectual property marks, I am tempted to think, another difference, one which enlivens the basic contrast. Property, and intellectual property, are concerned with controlling, or exercising power over, assets. The exercise of power is important in economics, but it enters only as a[n undesirable] sub-text. The ideal in the ethic of the discipline operates without the undue influence of any actor over another. Power, and its use, are deemed a spoiler, which enters with concerns about access, scale and dominance. By comparison, power is the direct stuff of concern among a wide set of the other disciplines — political science is the obvious case, with power as its subject and one of its centrepieces. In analyses which rest on social aggregates, power and its exercise are frequently instrumental for outcomes.

This chapter tries to tease out key points of contrast between the two seeming poles, or sets, of views. As the title "Social dimensions" suggests, the social aggregate (and its correlate, power) are the points of departure, as anchoring contrasts. Why focus just on points of contrast? These can be useful, it seems clear, only if they eventually suggest new agendas for research, or if they point to better policy. But the first step is understanding where frames of thought depart from each other. Only with a view of the other side is there a basis to begin to think about some approach in common, whether that is for research or eventually for policy. This chapter will try to look at each of the two sides from the perspective of the other.

Am I, as author of this piece, somehow free of the usual strictures on thought which a paradigm will impose on those who know it? Can I break away somehow from the usual human limitations and bring the "one, true objective word", so the reader now sees the light about *both* seeming poles — are some authors superhuman? Of course not. Inevitably, I will find myself blindsided by some preconception or another. But one can aim for an atmosphere which encourages open inspection of different approaches to similar or to the same phenomena, then one can look for new common ground while trying to preserve the advantages which the inevitable differences also bring. If this chapter takes a small step toward the first part — open inspection of different approaches — it will have met this author's aim.

The discussion below is organized around four topics in communications and information. Stimulated by the preceding sessions in the workshop, these four seem to bring out the contrasts between the two apparent poles of thought.

- *Innovation*. In economics, innovation is often thought the source for a better life. The other approaches sometimes share this view.
- *Productivity*. Productivity raises the prior question, what *is* the good life.

- *Governance*. Governance, as an issue, seems unavoidable in asking how to get there.
- *Universal service and intellectual property*. Finally, we can conclude with the two leadoff issues.

Where possible, the contrasts will be illustrated with examples from a National Information Infrastructure/NII, originally a US notion, and from the Global Information Infrastructure/GII, now a more widely shared idea. Unavoidably, a fair amount of discussion will refer to general precepts of economics and other disciplines, since it is tools of a discipline that are applied to the subject, communications and information.

Points of Contrast

Innovation

• *Social innovation*

Technical innovation has come to be appreciated as perhaps the prime source for productivity gains — the means by which living standards can be raised. Economics has devoted itself, *inter alia*, to incorporating technical change into our economic understanding. Technical change implicitly focuses on physical (and sometimes not-so-physical) delivery, and so implicitly focuses on the supply side. In other disciplines, in other approaches the *social* innovations on the demand side, which necessarily run in parallel with supply-side technical evolution,[3] take at least an equal part of intellectual attention.[4] The pathways of social innovation are, if anything, likely more complex than the technology from the world of man's tools, if only because so many individuals interact together, each of whom is individually enormously complex.

A small example comes from the virtual communities now taking form, in part along the Internet. It is said that there are cases which create an intimate group, but with as many as 900 members.[5] If this is true, a new human form has devolved in response to the technology. Along the way, there were probably intriguing modifications in personal and group modes, social innovations which may presage more far-reaching effects.

The contrast between a focus on social innovation or a focus on technical innovation is subtle, yet deep. Diffusion of innovation, as a field of study, occupies a position that is midway between economics and several other social disciplines; and it often has tried to maintain both foci together. Diffusion studies may, besides following the technology, also look at the process of adjustment whereby new technology is taken up by a group.

• *Technical innovation*

For technical innovation *per se*, a contrast is emerging, one which also reveals a worthy intellectual challenge. The contrast is perhaps best characterized by reference to "architectural innovation". This notion (at least in one form) suggests that a standard agreed for one area of technical activity enables innovation within the area. The standard is the framework for innovation at some subsidiary level — there is a sort of modularization of the change process.[6] An example in the US would be the Grand Alliance standard around HDTV, which acknowledged the importance of variety in implementations, to serve subsets of demand. In contrast, policy prescriptions for competition in an NII typically see standards as bringing homogeneity and so enabling competition. During a period of technical change different innovations are non-homogeneous and so, strictly speaking, not competing; standardization, in this view, brings homogeneity and allows competition.

The contrast in views is stark. One view sees further variety after standardization, the other sees the reverse, namely homogeneity. The intellectual challenge begins from the foundations of economics, where competition among homogeneous commodities is a primary building block. Economics has itself tried to incorporate variety into its models with Chamberlin's monopolistic competition and from Hotelling, for instance. The view that standard frames enable on-going variety, if it becomes salient, raises an intellectual challenge: how to conceptualize, with some analytic power, a world with sameness and difference in parallel. We will look closer at this theme, and challenge, when we turn to further aspects of social innovation, under "Governance" below. A workable notion of "openness" is at stake, among others.

• *Dynamics*

To portray the process of change has itself led to what is now a well-established contrast. Economics works from comparative statics, with a world view where stable equilibria are the successive anchors for any intervening change. An evolutionary economics has arisen, as a rump discipline, with a set of explorations intent on placing change itself at the centre of the conceptual stage. Co-evolution of the technical and social sides — the intertwined causality between them, rather than cross-impacts on price alone — becomes a focus. An ecological analysis, where co-evolution is the essential mechanism, has now taken some root in management strategy theory.[7] There the economic setting, and the system of operation posited to characterize that milieu, step from the background of the analysis into a prominent place.

Though evolutionary economics gets an increasing airing in the US, it is Europe, particularly from the Netherlands, Italy and Germany, where there are the most concentrated efforts. Beyond the confines of both sorts of economics, other disciplines and approaches may typically portray change as a feature central in their analysis. A co-evolutionary approach then helps to articulate an underlying concern, which also introduces the use of power. That question asks how a group — the users — may shape, and so may control, the technology which will in turn influence the underpinnings of the group.[8] Debates about the impact of violence when portrayed in mass media do go beyond the technology *per se*, but are a case in point nonetheless.

Dynamics imply the existence of a history. Though an historical approach to economics is now recognized, even with a Nobel, the paradigm has at its core an ahistorical bent. The notion of reversibility in demand perhaps best illustrates how the arrow of time points, if not both ways, then lightly at most in this framework. In a contrast which would counter that ahistoricity, path dependency has grown up as a line of argument — put strongly, this says "history matters". Unlike the evolutionary work, this line comes most from the US.[9] Other disciplines, beginning with history itself, would typically accord the past a privileged place in an inquiry.

With innovation a central feature of communication and information phenomena, certainly at the present, these several contrasts are one starting point for a dialog between those on separate sides of the seeming divide in views.

• *Method*

Not just with regard to innovation but generally, economics has evolved a methodology which is admired and emulated by a growing number of other disciplines. The emphasis on mathematical representation may divert attention from insights about the phenomena in question. More constricting, it may also close off use of key constructs which are not so susceptible to mathematizing. But the common mathematical language has supported very wide sharing of the economics world view. Even when mathematics is not used, one legacy is a relative rigour for analysis. Though rigour can threaten to be hidebound, by diverting or closing off attention, as said, it is also robust and gives strength to the understanding. We will return to this epistemological balance below, under "Governance".

Productivity

Concern for productivity, and particularly productivity increases, may centre on the individual, firm, industry or a whole society. At the broadest level, the question arises, to produce what — what is good, what is valued?

• *Value or price*

Adam Smith's original "paradox of value" brings forth a hoary contrast, one often papered over but still at the root of seminal, continuing differences. This paradox asked why some most valuable items, such as the air we breathe, had negligible or no price. Economics eventually resolved the quandary by reference to the inexhaustible supply of air[10] in a balance with demand. In contrasting approaches, however, money prices may be shown not to match the valuation found. Prices may not fit well with an analysis of valuation from marketing or psychology, for instance. Communications and information are particularly sensitive to the problem. Examples are flat-rate pricing for telephone, and the notorious question of valuation for information. In response, economics has tried to find hedonic prices.[11]

At a deeper level, the contrast centres on the process whereby value emerges and changes. Though the balance between supply and demand in economics is in some sense social, price is the primary point of contact among individuals. A sociologist or an ethicist in philosophy, by contrast, is likely to be concerned with broader and richer dimensions of social interaction. Value will there be seen as created, and evolving, through social mechanisms.

• *Development*

As a pivotal product growing from five centuries of the Western Enlightenment, economics embodies a key thrust of the era, perhaps subliminally but nonetheless. Put simply, developing societies may, through a variety of passages, achieve the economic status of, and be admitted to, the industrialized [productive] brotherhood. Pareto's optimum, where advance for one is not to disadvantage another, is one distillation of an ethic for improvement.

The question, and contrasts, broached just above, "what is the good?" sit palpable here. Beyond that, the contrast between an individual-based versus a social approach to analysis begins to take on detail. A central facet of development is learning. Though this is learning between societies and so is at a macro level, learning may be taken as an essentially interdependent activity and thus as emphasizing the social interaction.

The typical agreement to build-operate-transfer a modern communications infrastructure in a developing country (perhaps for GII) aims to inculcate the technical capabilities locally. When brought down from a lofty macro level, to microeconomic events, the learning that is required can appear as a fairly complex social event. That emphasizes the interdependence.

To reach everyday microeconomics is a small step. If we are all developing, all the time, learning is endemic. Economics recognized learning-by-doing, with roots in self-learning. Alternative approaches, by contrast, see the interdependence. The analysis starts as multi-party and social; to account for the interdependence, the story about learning also needs to be dynamic. Questions about "system productivity" may apply in this style of dynamic analysis. Competence could be characterized as the stock which results from the learning flow.[12] Education as a discipline is the first source for many alternative ideas about learning.

The final contrast here engages the larger question directly, what is the "good"? Built into the notion of diffusion is a directionality to the flow of ideas. "Better" technology has been validated someplace else, and now it can be taken up by others. Those who are beyond a diffusion approach, including a strict interpretation of individual preferences inside economics itself, may broach an alternative. Though put in many different ways, the position is essentially social. It envisions what amounts to a negotiation among camps, an exchange in which values from both sides are offered for consideration. From a "traditional" society the offer might be "our style of family values". ["Can your new technology network take full advantage of them?"] is the implied question.

The contrast here is between a uni-directional and a bi-directional exchange, about values. The potential adopter may note that communications technologies can be used to hinder as well as to help[13] (in one scenario, over-reliance on telephone communications diminishes the quality of personal relations that would otherwise make better organizations). Of course there may be no exchange at all, except by force: a "holy war" intends to preserve or spread a set of values. But in the full two-way exchange, values from traditional societies may also influence outcomes in the more industrialized. Then we are all developing, all the time.

Governance

We look first at the structure, or organization, to be governed.

• *Structure/organization*

Though the question of structure is an on-going topic in economics (as are the rest), the divide between markets and hierarchies has suggested a bright line clearly distinguishing these two. In this world externalities are a main concern.

From other points of view, notably evolutionary economists, sociologists and management scientists, intermediate forms are prevalent. These scholars see some porosity and permeability in organizational boundaries. When looked at across time, dynamically, the elements may both endogenize and exogenize, with some alternation. At some times the elements join more together, at others they are more separate. The result is a more nearly organic entity, where social linkages play a main role and where structural/organizational form alternates, depending on relative strengths between independence and interdependence.

How to characterize the elements, one relative to the other? When they are together, they are in some way apparently the same; but when separate, supposedly different. This is a more interesting formulation of the intellectual question encountered above, around standards and variety: how to encompass sameness and difference, *in the same set*.

The issue is especially salient for communications, where networks from separate carriers must interconnect, to reach a wider group. While the individual carriers remain distinct, essential similarities of some sort must make interconnection possible, and greater interoperability drives greater similarity. Thus merge the standards and the organizational forms of the intellectual question.

If taken from the point of view of users, in interconnected groups, this becomes the issue of the public as against the private. In a dynamic world, an individual user may move mentally among affiliation in a larger, a medium or a smaller group. A subject that is considered appropriate for airing in the medium group may be too private for the larger group, but perhaps taken to be public knowledge by the smaller group. An approach that deals with structure in this dynamic way puts the analysis of the public and of privacy in a new light. The question of privacy, and also the question of what is public, are another area which is particularly salient in communications and information.

The contrast between a bright organizational line and a dynamic approach with (in shorthand) "social porosity" has special relevance for the study of communications and information. The matter will be revisited yet again under the subsection below on power itself.

Before we leave this section, we notice that the obverse of the last paragraph is also so. Communications and information play a special part in either of the conceptualizations of economic (and other) behaviour. In economics, that point has been the thrust overall of information economics. In the alternative dynamic approach, communications may be the linchpin. In one formulation of the alternative, knowledge is a shared construction. Communications links make possible a more complete, more robust shared picture. The dynamics across time involve shifts in frame of reference, from one accepted view to a next.[14]

That is, of course, the over-arching topic of this chapter on "Social Dimensions": the competing frames of reference for communications and information.[15] Discussion in the hallways, during the workshop, brought out a basic contrast between the systems of thought being considered. One might be construed as "tightly" structured and axiom based; the other, "looser" and more open. From the dynamic approach just above, one description of shifts in frames of reference alternates between a tight, agreed frame and a looser frame, which allows the experimentation necessary to establish the next tight frame. This is a dynamic formulation of the epistemological balance discussed earlier, regarding methodology in economics.

• *Power*

We turn to contrasts at the heart of governance, around the use of power, itself.

Perhaps it is an irony that economics, the discipline for which power is a spoiler, has built in constructs where power is implicitly recognized (or maybe that is a natural response, to incorporate the phenomena). Hierarchy, as against market, and principal/agent — hierarchy writ small in other words, hierarchy on a modular basis as it were — describe power relations, but implicitly.

In contrast, the alternative formulations may feature the use of power. Social approaches have, above, filled out with interdependence for learning, portrayed dynamically; with power included, they take on full flower. This exercise of power takes the more complex form which accommodates porous organization boundaries and alternations in structure. That is the power of hierarchy when elements are together, but it also encompasses the elements separated, elements which may yet rejoin again. Along with the power of individual actors, there is an equal role for the power of shared agreements about social organization — the values, or culture, that might be the subject for a study of community, in philosophy, anthropology or sociology.

These contrasting approaches set different terms for access. In the alternative approach, access occurs along the time stream, as a function of periods when organizational elements are more together, more hierarchically layered and more coordinated. Openness, for a technical architecture, would depend on this dynamic formulation.

At some fundamental level, the contrast between the two poles centres on the primitive response to power and its exercise. For economics, and here for a range of other disciplines, the response is either to break the grip of power, or to corral and control it — anti-trust or regulation. For the contrasting view, perhaps the tension is with Western Enlightenment styles of thinking. Hajime Oniki, during the workshop, described for us how the notion of regulation has required a loan word to be brought into the Japanese language, from outside. In that society the Western notion of regulation is so much a part of life — it is "in the air" — there is not a native word, as such.

That could suggest an alternative response to power, one which attempts to harness and to use power productively. But similar cases may also be found in the West, without particular intellectual articulation though. Sweden, and perhaps Scandinavia generally, seem in their dominant telephone operators successfully to have harnessed the power of large organizations. Sweden for instance enjoys both superb service and low prices.

• *Policy*

The discussion of power has already broached policy, really; now we turn to mark contrasts in two closer-up topics: competition and pricing. When we move to policy, the contrasts shift from the intellectual arena of disciplines to wider constituencies, as already seen in the last section.

Competition is now a very widely accepted prescription, throughout the world. In the contrast, social dynamics also identify the role of consensus and cooperation. Though not presently so vocal, these alternative prescriptions also maintain some adherents around the world.[16]

For pricing, the debate inside economics itself continues almost unabated. But outlines are visible. Though price caps have enjoyed and do enjoy wide support, with efficient component pricing the dialog itself is returning to the regime of caps and floors which structured the debate around the breakup of AT&T. In the intra-disciplinary debate, price is again being predicated on cost,[17] which returns to a main dictate of the paradigm.

By contrast, a range of other voices would favour what could be described as community pricing. This contrast in fact stretches back through the history of policy for telecommunications. National administrations long continue to insist on "low" pricing, intended to bring all onto the net. "Value of service" pricing encapsulated the idea, and it has been used even by private carriers. Today, pricing the Internet is one of the hottest points of contention for GII policy. There the alternative view would maintain some features of the present Internet practice, in which the subscribing institution acts as surrogate for the community within that institution. A key effect is pricing for that community, rather than some cost basis on individuals or usage.

The individual and the group

Because communications (and information implicitly) invoke physical networks, it is not surprising after all that alternative analysis concerned with social groups and human networks may come to some notice. This chapter has sought to outline contrasts between two apparent poles of thought, where many of the contrasts depend on a social versus an individual approach. We conclude by returning to our two "polar" subjects, universal service and intellectual property. As we do, I move from being observer, reporter and scribe (however far from "objectivity") to speaking as protagonist in the dialog.

On the one side, the canons of intellectual property help crystallize, I believe, the place of the individual in this analytic drama. If knowledge is really the product of shared effort, to parcel out a piece to an individual does damage to the shared heritage. This will even make a difference, when later the piece is needed by others, to assemble the *next* piece, along the path of knowledge. That which is necessarily shared, and is needed for later sharing, has been artificially hewn to a single individual. Yet, there are compelling reasons. The place of the individual in the process cannot and must not be denied. The individual is the only real starting place for any later, shared — productive — results.

On the other side, universal service is that paean to the social group — to groups which pretty clearly are fixtures in human outcomes. But history is strewn with the figurative and literal corpses of those run over by a bulldozer in the hands of some demagogue who acted "on behalf of the group". Rather than the prospect of sharing, reliance on a group may augur grim, or grisly results. Yet the sharing is necessary for many, if not most, desirable outcomes. A very large number of achievements are realized only through concerted effort.

Between the two, I will suggest an interpretation. If we imagine that the place of the individual, as against the sway of some group, has been very hard won, across history, then we appreciate how important is a style of analysis which does not do damage to the individual's position. If we also conclude that man is a social animal where outcomes simply cannot be divorced from affiliative memberships, then we may see an analytic course, over time, to take. Having opted to preserve the place of the individual, we may then try to locate that individual in a social setting which does some justice to the irreducibles of those social facts.

This chapter has tried to portray some contrasts between styles of thought which may sit, to some extent, on separate sides of this divide. If it meets my aim at all, those contrasts may suggest points where both sides can engage the questions implicit in such an agenda.

Notes

1. Macro views of economics, by their second-generation status in the discipline, illustrate how the individual sits at the original cornerstone of economics thinking.
2. This, despite repeated turns, for other purposes, to the individual as the unit.
3. See Bruce Mazlish, *The Fourth Discontinuity: The Co-Evolution of Humans and Machines* (Yale, 1993), for the definitive discussion.
4. The breadth of turf calls for a variety of outlooks and needs references seminal for the several disciplines. The aim, of course, is those pieces which best represent the points of contrast discussed here, on both, or the several, sides (not an exhaustive compendium, which as far as I can see would go too far, given the number of disciplines).
5. Lee Sproull, *Connections: New Ways of Working in the Network Organization* (MIT).

6. Where workshop participants were instrumental in bringing a point to the attention of the meeting, they will be duly noted. In this case, Peter Hagström discussed architectural innovations. My own previous work (without referring to "architectural innovations") has sought the policy implications. See for instance Allen, 1991.
7. See for instance James F. Moore, "Predators and Prey: A New Ecology of Competition", *Harvard Business Review*, June 1993, pp.75-86 (a book is forthcoming).
8. Deanna Campbell Robinson noted the question of users trying to shape their technology.
9. Most identified with Paul David and a set of following authors; see for instance Paul David, "Path-dependence and Predictability in Dynamic Systems with Local Network Externalities: A Paradigm for Historic Economics" ...
10. Then air pollution came back to the question.
11. Hajime Oniki and Cristiano Antonelli both discussed aspects of these questions.
12. Gunnar Eliasson reminded us about the role of competence.
13. Don Lamberton reminded us of Veblen's point.
14. With Thomas Kuhn, The Structure of Scientific Revolutions, the original description.
15. Michel Menou pointed out that both ethnography and psychotherapy are disciplines well suited to help understand underlying structures of thought.
16. Don Lamberton reminded us how there have been periodic swings toward, and away from, competition as the policy prescription, in the intellectual arena.
17. Cristiano Antonelli sketched this development for the workshop.

CHAPTER 10

Comment: Ann Moyal

The foregoing analysis is an elegant summary of the two polar approaches to telecommunications research — economics and other disciplinary. But the very tightness of its argument and its concentration on the differing "styles of thought" arches too broadly over that spectrum of research viewed by some members of both streams as central to an understanding of the social dimensions of telecommunication in demand, pricing, regulation, and policy.

Allen focuses the place of the individual versus the social group, and suggests that questions in respect of the latter might be the subject for a study of community in philosophy, anthropology or sociology. But as Alleman and de Fontenay (chapter 3) point out, the telecommunications market is in a state of turmoil. We face a world where services are being constantly developed, and new alliances, mergers, and joint ventures consume our telecommunications organizations and precipitate telecommunication legislation. While we know little enough about human needs and uses where "a voice call was a voice call", we face an increased necessity and challenge to understand communication behaviour in all its new diversity.

Perhaps, as Taylor (1994) asserts, we have, internationally, developed a greater knowledge of residential "access". Yet how scant is our knowledge of residential and social usage. As sociologist, Aronson, observed as early as 1971 in respect of POTS: "The questions concerning telephone use yet to be answered are who talks to whom, for how long, and for what reasons and with what results." Or as Nobert Wiener, still earlier, affirmed: "society can only be understood through a study of the messages and the communication facilities which belong to it" (1950:18). Such questions still remain only most sketchily answered, while other questions, cross-cultural, international, cross-media, gender — to name a few — encrust an increasingly complex consumer scene.

Alleman and de Fontenay in discussing demand, observe the necessity of understanding human behaviour — "cultural and psycho-social" — as an integral part of the general desire in understanding communication behaviour, and find a key to this in such questions as whether women and men use the phone differently, the kinds of communication which are transmitted, their information content, and whether people-to-machine communications are alienating.

Menou too encompasses this view in his judgement that "the analysis of demand" must proceed from the knowledge of the underlying utility of the communications. The first component is the information itself: "what goes into the pipes". There is consensus on likely changes over time in the utility function of consumers, as humans learn to appreciate new goods and services, and on the need for a research agenda that takes account of this.

Alleman and de Fontenay further note, and the point is central to Moyal's (1989, 1992) and Rakow's (1992) that gender differences will become increasingly significant with the increasingly diversified telecommunications good in an environment which is ever more competitive. Significant variations will clearly occur between different nations (industrial, developing, small and large geographic, historically conditioned) and cross-cultural research is essential here. Already in international voice calling, both business and social, gender differences have been observed in Australia, and broader implications noted.

Universal service

Allen has characterized "universal service" as "the fundamental" which separates the two conflicting disciplinary points of view: viz. the market in economics is concerned with "penetration up" to some point, while the sociology/communication's concern for the welfare of a group, "counts down", and asks "are all members accounted for?" Such differences, however, are not mutually excluding. In Australia, and no doubt elsewhere, there is an arena of community-based, advocacy and small group research which is demanding a place for its discourse and knowledge content alongside the more pervasive carrier and technology-push discourse, within the "policy moment". This discourse extends the notion of universal service which Cave (chapter 8), for example, defines as "basic telephony services to any customer who requires it, at an acceptable or affordable price", to a concept of equitable access to new telecommunication/information services as an intrinsic part of the "social good".

Rigour of analysis is not at present a conspicuous feature of this community-based research; but its participation in contributing to the base of information assumes importance as carrier data becomes closed to researchers. There is an argument for some mentoring by the research community, a process seen as desirable by the groups themselves, to introduce more rigorous methodologies and aims.

Cross-cultural studies of national institutional and organizational structures should form an important part of the research agenda. How does history modify different modes of operation? Can/should it serve in national policy decision-making? (see Moyal, chapter 11). Sweden, and perhaps Scandinavia with dominant telephone operators, have reportedly successfully harnessed the power of large organizations, monopoly-driven Sweden offering remarkable service and low prices. Comparative research will be enlightening, and, it is hoped, timely.

Policy

Competition, USOs, constituency and group interest figure as part of the policy research evaluation. A central problem has been, and is, the speed of legislative action before adequate research is available. A CIRCIT conference in Melbourne in 1991 evoked the memorable comment from a senior telecommunication policy bureaucrat, "We haven't time for the research; we must get on with the legislation!" The dominance of the economic paradigm in policy analysis is marked. Bureau of Transport and Communications Economics research studies are highly focused. Research on the social dimensions of telecommunication receive scant aid from government. Choices, however, are being made that lock in society and cancel alternatives without the attendant process of scholarly or community input.

There is a context in which social dimensions need to be directly addressed. There is an evident need, given the pace of telecommunications advance, for the encouragement of national, and collaborative international research communities engaged, and supported, in work on multidisciplinary aspects of the impact of telecommunications, contemporary, historical and changing. Lamberton's point on research agendas (1995:371) is pertinent here: "If we agree that the *raison d'être* of telecommunications is not simply the result of the availability of technology and industry supply but their adequacy in meeting needs of individuals, groups and organizations to communicate, then social dimensions should have priority in research. The major issue is whether the provision of information and information facilities does or does not make a difference in abilities to cope with problems and in what way do telecommunications affect power structures within society."

Nightingale (chapter 6) offers a further thought on the pace of change in the policy and regulatory field. "The very phenomena of rapid change", he notes, "may carry with it consequences of resistance, distrust and political disruption, as the apparent losers fight to defend their past benefits. A slower pace of change, and an adequate consideration of distributional issues ... may well create a more politically sustainable market structure in a democratic agenda, perhaps not mainly from economists."

References

Aronson, S., 1971, The sociology of the telephone, *International Journal of Comparative Studies* 12(3), in: C. Gumpery and R. Cathcart, eds, *Intermedia*, 3rd ed. (Oxford University Press, New York).

Kellerman, A., 1992, The social demand for international telecommunications: The hidden promise, *Telegeography 1992*, IIC:12-15.

Lamberton, D.M., 1995, Telecommunications economics: research agenda, in: Bureau of Transport and Communications Economics, *Communications Research Forum 1994*, Vol.2, BTCE, Canberra, 363-376.

Moyal, A., 1989, The feminine culture of the telephone: people, patterns and policy, *Prometheus* 7(1):5-31.

Moyal, A., 1992, The gendered use of the telephone: an Australian case study, *Media, Culture and Society* 14:51-72.

Rakow, L.F., 1992, *Gender on the Line. Women, the Telephone, and Community Life* (University of Illinois Press, Chicago).

Taylor, L.D. 1994, *Telecommunications Demand in Theory and Practice* (Kluwer Academic Publishers, Dordrecht).

Wiener, N., 1950, *The Human Use of Human Beings: Cybernetics and Society* (Sphere Books, London).

CHAPTER 10

Comment: Richard Joseph

Allen reports on a tension which developed in the Workshop between two disciplinary frameworks. He has nominated the first disciplinary framework "economics" and the other "an array of other disciplines" which includes fields such as sociology and political science. In order to throw some light on this tension David Allen has written a most interesting and provocative piece attempting to highlight key areas of friction. He has provided comments in four significant areas for telecommunications: innovation; productivity; governance; and universal service and intellectual property. The approach adopted by Allen is useful in identifying a whole range of research questions. Researchers and policy-makers work using different frameworks. The results of their work will be coloured by these intellectual frameworks and Allen is very astute in focusing in on the points of friction. Allen works with the idea of polarity (although he points out correctly that it would be wrong to be too dogmatic about these poles) and areas of overlap to make his case. I would like to suggest some thoughts on possible research options that I hope will bridge the gap that Allen has identified.

Economics and politics

While it is possible to see economics as the dominant paradigm for interpreting telecommunications and social and policy issues, I think a fruitful path is not through marrying the differences between the two areas but by exploiting an area of commonality. By this I mean seeing the economic language of telecommunications policy as a form of political rhetoric. Instead of being an "objective" measure for assisting policy advice, economics becomes the very "glue" which cements political and institutional interests in place. This requires a more circumspect look at the implicit political foundations of economics together with an understanding of the role of political language and rhetoric. Nelson's work on the doctrinal aspects of private enterprise would be a starting point.[1] McCloskey's work on the rhetoric of economics would be another.[2] I think we can develop a perspective to see economics not only as a handmaiden of political forces that influence telecommunications policy but as a vehicle for structurally favouring certain political interests. Some examples which come to mind are the use of the concept of market failure in the allocation of resources to science and technology[3] and the use of particular accounting methods to determine universal service costs.

Technological innovation and social innovation

In his discussion on innovation, Allen discusses the study of the diffusion of technology as a bridge between what he calls "technological innovation" and "social innovation". As far a telecommunications research is concerned, there is need for the sociologists of technology, historians of technology and analysts of large technical systems to devote their attention seriously to telecommunications in general. A useful related area of study would be the Internet. By intellectually separating and categorizing innovation as "technological" or "social" one simplifies technology as an "artefact". Technology is more complicated than this complexity needs to be incorporated as much into social theorizing as society needs to be built into histories to technological change. It is important to recognize that technology structures human behaviour and it is in turn shaped by human values. Likewise, technology enables certain actors to pursue their objectives. Technology and related systems are also highly symbolic and carry political meanings. So instead of seeing the Internet as a creation of new technology and new social groups forming as the technology has an impact, I would argue that much can be learnt from history, especially from the evolution of large technical systems.[4] There is an urgent need for reconceptualizing telecommunications innovation in a more holistic and historical approach.[5]

Governance

Allen has raised a number of issues under the general heading "governance" which provides the basis for further research. My remarks relate to the notions of deregulation and liberalization which are dominant in telecommunications at present.

First, if telecommunication reform is to be understood politically then telecommunications needs to be more properly located in mainstream political analysis. Since the state has played a central role in the telecommunications industry through sanctioning monopoly rights, it seems logical that theories of the state and governance would be helpful. Vincent Mosco's work has attempted to forge this link and I believe that it is useful in transcending existing industry interpretations of deregulation as due to forces such as "technology", "demand", and "globalization".[6] Of course, national comparative case studies should provide some empirical base for these theoretical propositions. The international dimension in telecommunications should not be overlooked either. Existing models of globalization in telecommunications are inadequate for capturing the political economy of the emerging international telecommunications regulatory system and there is an urgent need for work in this area.

Second, Allen mentions the balance between public and private in telecommunications. This is very topical as countries deregulate and seek solutions to investment problems through privatization. Economics has much to say about the benefits of private over public organization in telecommunications but I would like to see these contemporary debates tempered by more historical analyses. If this were so, I believe that the dogmatism expressed by some economists could be watered-down. Arguments for private ownership in telecommunications may be seen to be rhetorically based in order to achieve political objectives other than those predicted by the economists. Indeed, if we are going to change structures to a more market-oriented base, there seems to have been little work actively incorporating the consumer movement into these structures. I believe that formulating a framework that legitimizes formal consumer representation that actually empowers residential consumers is vital in shaping a communications future that is at least partly humane.[7]

Third, Allen talks about power as being at the heart of governance and discusses the role of shared agreements about social organization in exercising this power. In telecommunications, the concept of deregulation is one such "shared agreement" in need of further analysis. Its political dimension needs to be outlined. The work by Horwitz[8] and Mosco[9] are valuable in this regard. Mosco uses the term myth to describe deregulation and this is a useful approach. Robert Babe's work on analysing the "mythologies of Canadian telecommunications" is an application of this form of analysis to a national case study.[10]

Finally, Allen points to the pricing of the Internet as an important area of further study. The Internet has grown so rapidly and has captured the imagination of so many people that in itself, it would be a fruitful subject of research. Of most value here, I believe, is an analysis that incorporates the "social shaping" of a large technical system. History is again vitally important here and we can learn from similar network examples in the past. However, the pricing debates which are occurring now represent decisions which will determine the nature of the Internet in the future. Our inability to get the Internet into some sort of historical perspective could mean that its liberating aspects are stifled by those whose interests are best served by powerfully utilizing both technology, organisational structures and economic rhetoric to get their way.

Notes

1. R.R. Nelson, 1981, Assessing private enterprise: an exegesis of tangled doctrine, *Bell Journal of Economics*, 12(1).
2. D.N. McCloskey, 1985, *The Rhetoric of Economics* (University of Wisconsin Press, Madison).
3. R.A. Joseph and R. Johnston, 1985, Market failure and government support for science and technology: economic theory versus political practice, *Prometheus* 3(1):138-155.
4. J. Summerton (ed.),1994, *Changing Large Technical Systems* (Westview Press, Boulder, CO).
5. S.J. Becker, 1993, The political economy of early telephony in Germany, in: J. Wasco, V. Mosco and M. Pendakur, eds, *Illuminating the Blindsports: Essays Honoring Dallas W. Smythe* (Ablex, Norwood, NJ).
6. V. Mosco, 1988, Toward a theory of the state and telecommunications policy, *Journal of Communication* 38(1):107-124.
7. W. Streek and P. Schmitter, 1985, Community, market, state — and association?: The prospective contribution of interest governance to social order, in: W. Streek and P. Schmitter, *Private Interest Government: Beyond Market and State* (Sage Publications, Beverly Hills, CA).
8. R.B. Horwitz, 1986, Understanding deregulation, *Theory and Society*, 12:139-174.
9. V. Mosco, 1990, The mythology of telecommunications deregulation, *Journal of Communication* 40(1):26-49.
10. R.E. Babe, 1990, *Telecommunications in Canada: Technology, Industry and Government* (University of Toronto Press, Toronto, Canada).

CHAPTER 11

Interdisciplinary research: history

Ann Moyal

What does the historian have to offer the debate on telecommunications, its economic development, its organizational structures, its regulation, its regional and national perspectives, and its social implications? The answer is clear. History illuminates contemporary concerns.

The global ethos dominating telecommunication development over the past decade, the thrust of competition, new technology, interconnection, proliferating services, and the high public promotion of an "information superhighway", have all tended to lift the focus from national identity in telecommunication and to present a potent sense of future that has little connection with the past. There has, resultingly, been scant encouragement to scrutinize and consider the social and political factors that have forged distinctive national telecommunication environments and governed particular responses and adaptations to the new. Specific questions of the diffusion of telecommunication innovations in the historical context have been left largely unexplored.

Even before this quickened pace of change, only a moderate stream of historical studies had emerged to throw a lantern light on telecommunication administration and organization, the spread of its technologies, and the impact of telecommunication in shaping social change. It is not my purpose to survey these studies, many written as pamphlets and journal articles, although substantial examples such as the International Telecommunications Union, *From Semaphore to Satellite* (Geneva, 1965); John Brooks, *Telephone: The First Hundred Years* (Harper & Row, NY, 1976); I. de Sola Pool (ed.), *The Social Impact of the Telephone* (MIT Press, Cambridge, MA, 1977) and James R. Beniger, *The Control Revolution: Technological and Economic Origins of the Information Society* (Harvard University Press, 1986) stand as foundation works.

A sense of history has, in addition, fired several major telecommunication industries to evaluate their past. Peter Young's *Power of Speech: A History of Standard Telephones and Cables 1883-1983* (George Allen & Unwin, London, 1983), examines the reversals and triumphs of a company originally established as a British agency of American Western Electric across a century of activities in cable-laying and early broadcasting, its organizational and business progress, the internal climate of innovation that spurred such revolutionary STC inventions as pulse code modulation and optical fibre communication, and the influence of customers and governments. Even more strikingly, the multi-authored *L.M. Ericcson 100 Years* (L.M. Ericcson, Sweden, 1977) edited in three volumes by Arthur Attman *et al.*, provides a detailed and comprehensive view of the far-reaching coils of the Ericcson telecommunication octopus, whose chapter on "Three key markets, Mexico, Australia, Brazil" in Volume II covering the period 1932-1976, shed invaluable light for me on Ericsson's approaches and penetration in Australia including its part in Telecom Australia's major systems developments. The usefulness of these works to business and corporation history and to the history of telecommunication cannot be over-stressed.

In the field of national telecommunication histories, however, only a very small cluster of studies has, as yet, appeared. Canada set the scene with two works, Robert Collins, *A Voice from Afar: The History of Telecommunications in Canada* (McGraw Hill Ryerson, Toronto, 1977) and E. Bernard, *Long Distance Feeling* (New Books, Vancouver, 1982), the first of which — a racy, popular story geared for those whose concentration span was short — served scant evaluative purpose. My large social, political and technological history, *Clear Across Australia: A History of Telecommunications* (Nelson Australia, Melbourne) followed in 1984. Edward Harcourt, a senior official of the Overseas Telecommunications Commission (OTC), published *Taming the Tyrant: The First One Hundred Years of Australia's International Communication Services* (Allen & Unwin, Sydney) in 1987, and in 1994 A.C. Wilson filled out the Australasian picture with his slim historical analysis, *Wire & Wireless. A History of Telecommunications in New Zealand 1860-1987* (The Dunmore Press, Palmerston North). Michele Martin's *"Hullo Central": Gender, Technology, and Culture in the Formation of Telephone Systems* (McGill-Queen's University Press, Montreal), with its Canadian focus and ranging bibliography, is an important addition to the genre.

Such histories vary in content and style, but Harcourt, Moyal and Wilson's are scholarly evaluative works. *Clear Across Australia* has been hailed as a pioneering model in the national historical field. Commissioned by Telecom Australia in 1980 and prepared under the watching eye of a distinguished external Editorial Board (with one Telecom representative) appointed to ensure authorial independence and a critical and comprehensive view, this 437 page history carried the narrative of communication development from the introduction of the telegraph in 1854, through the growth of the separate

Colonial administrations, the federation of services under the Commonwealth of Australia in 1901, to the creation of Australia Post and Telecom Australia as separate statutory bodies in 1975, and into the growing pressures for privatization in the early 1980s. At the same time it examined the evolution of the contributing professions of linesmen, telegraphists, telephonists, technicians, engineers, and their unions; explored the administrative choices and the transfer across 180 years of diverse technologies; evaluated the influence of political parties, key politicians and directors-general in a government-owned telecommunication monopoly; examined Australian telecommunication research; unravelled the motivations of Australia's satellite policy; analysed the rapid economic and organizational rise of Telecom Australia from 1975; assessed the adoption of new automative technologies in the seventies and eighties and their critical industrial implications, and probed the social influence of telecommunication in a huge and distant country.

Written from archival, parliamentary, and printed sources, and wide oral history interviews, the book served several purposes. Within Australia, it formed a foundation study in a neglected arena of social, organizational, technological and business history, serving researchers in multidisciplinary fields. It became an informing source for a changing flow of Telecom administrators, staff and Board and Ministers of Communications. It served a similar use for staff of the new regulatory body AUSTEL and the subsequent Telecommunications Ombudsman's Office, as it did for the Australian Telecommunications Users Group (ATUG) and members of the various telecommunication and broadcasting advisory bodies that peppered the Australian landscape from the 1980s. It also fed into related fields of administrative, organizational, union and industry studies and telecommunication law. It formed, too, a landmark study in the history of Australian technology and science.

Reading it now, I am astonished (as authors often later are) by the enormous range of material covered and the critical cogency of the book. Even so, and at the time of writing, I was conscious that each chapter opened up themes and questions for a crop of PhDs. Nightingale has sighted one in his desire for "a thorough investigation" of the "two supplier" policy adopted by the PMG's Department in the post-war period as a form of "integration" (chapter 6). While economic historians and economists have been slow to take up such questions, one of the undoubted merits of pioneering national histories is to provide a key to what we do not know and to enlist the interest of other researchers.

Nonetheless, with its gaps and pointers to further research, the book's overriding value remains (i) nationally, in the illumination it provides to telecommunication management and to multidisciplinary scholars on the challenges, adaptations, innovative thinking, achievements and mistaken courses that lay at the core of the largest government institution in Australia (the Postmaster-General's Department from 1901-75) and the subsequent Telecom monopoly, and (ii) cross-culturally, in the pertinence of the experience of a developing industrial country of small population and significant distances to other national studies of telecommunication.

In reviewing *Clear Across Australia*, Marginson (1986), a former senior member of the PMG's staff and, at the time, Vice-Principal of the University of Melbourne, gave a verdict that has resonance for other countries. "It is", he wrote, "for those whose life has been involved in the public sector, very satisfying to see the emergence of some scholarly and professional works dealing with those major public corporations that have shaped our society. Ann Moyal has provided us with a detailed account of an area that has had little quality documentation to date ..."

> Whether statutory corporation or public company, there is a very real obligation on the controlling board of each major institution to see that its key records are preserved. More important there is a need to ensure that a comprehensive and analytical study of its development is produced. The production of such a study is not inexpensive, and the work should not be attempted by amateurs, however able. It should also be pursued, if possible, while there are sufficient old hands available to assist in illuminating the factual history and assisting the interpretation.
>
> A great deal emerges from a book like this, particularly when it is written with a perception and understanding of the patterns of social history ... There have, moreover, been few occasions where a near-contemporary account has the feel of an impartial summation.

The research agenda urgently requires further national histories of telecommunication development. Wilson's New Zealand venture *Wire and Wireless*, a work of independent scholarship conducted across ten years, coheres around four themes; the domestic factors that shaped the country's telecommunication until the separate corporations of New Zealand Post Office and New Zealand Telecom in 1987; the imperatives generated by equipment and technologies; the impact of specific politicians and senior department staff; and less satisfactorily, New Zealand in an international context. The author usefully returns to the theme of the importation and adaptation of new overseas technologies into the local environment which provides some counterpoint to the larger compass of Australian experience. Harcourt's explicit focus in *Taming the Tyranny* on Australia's links with international cables and later satellite services further confirms the relevance of history in its analysis of *post hoc* decision-making, forward planning, and of events fiercely shaped by competitive offers and political responses.

Historians may work alone, but history is a collaborative enterprise. The micronarratives belong within the macronarrative of history. We have much to learn sociologically from the distinctive telecommunication experiences of countries such as Japan, France, Italy, Sweden, Russia, and China, for example, where different geographies, concentrations of population, comparative technological sophistication, and varying nearness and distance form a contrasting part, as we have from such telecommunications giants in North America, and the developing regions of the globe. These experiences and the light they shed on human behaviour, stability, and needs have contemporary relevance as we move towards globalization and the possible social disorientation inherent in change in telecommunication processes in too short a time.

The Australian evidence on women on the telephone and the highly supportive feminine culture of the telephone that has grown up to bridge distance and provide a vital kin and social support system in Australia (Moyal 1989, 1992) is a potent example of the shaping of a national culture through the long existence of the untimed local call. Despite duopoly and the prospect of wholesale deregulation, this issue continues to influence and modify Australia's telecommunication policy and to allow a social policy approach both to USOs and untimed local calls through cross-subsidy and other structures.

History, hence, does modify different modes of national operation, and can contribute perspective to national policy-making and national goals. Certainly the place of gender in telecommunication, and its historical, contemporary and future relevance, e.g. the role of women as distinctive, influential customers and consumers stands as a key area for cross-cultural research.

Nightingale (chapter 6) also signals areas where historical analysis will serve contemporary affairs. He pertinently assumes that "history will doubtless reveal more complexity than is now apparent in what appears from present perspectives to be a remarkably simple environment for the PTT" and adds, "The research agenda must include such an analysis if the present is to be well understood". More broadly, he asks in respect of the conventionally accepted view that "without the cost saving improvements in electronics, micro- and opto- included, few of the present or contemplated developments in telecommunications would have been possible", is this view supported by history? A further central question for examination is "to what extent these technological changes were beyond the control of the telecommunications sector, exogenous and imposed by force of technological possibility for cost saving and service innovation?"

The long arm of history, and vigorous historical analysis, appears vital to our understanding and our critical participation in shaping technological and market "advances" ahead. The multidisciplinary need is acute. This Workshop has revealed considerable soul-searching on the part of the economists; demand and its complexities has been placed under the microscope. But the turbulence of new markets relates to many underlying concepts where knowledge of earlier experiences and predictions are central to the evaluative task. I will echo John Nightingale: "Understanding these phenomena is a major task. It should not be left to economists."

References

Marginson, R., 1986, Review of *Clear Across Australia, Historical Records of Australian Science* 6(3).

Moyal, A., 1989, The feminine culture of the telephone: people, patterns, and policy, *Prometheus* 7(1):5-31.

Moyal, A., 1992, The gendered use of the telephone, an Australian case study, *Media, Culture and Society* 14:51-72.

CHAPTER 12

The contribution of information science to telecommunications research

Joyce Kirk

Information is the central phenomenon of concern in information science, described as "the science that investigates the properties and behaviour of information, the forces governing the flow of information, and the means of processing information for maximum accessibility and useability" (Shera and Cleveland, 1977:265). For this reason, information science has a unique contribution to make to the study of telecommunications and information activities. It has the potential to enrich the study of phenomena in telecommunications as well as provide a context for the study of these phenomena. In terms of the structure of the Workshop the areas of research most likely to be informed by information science are those of

demand studies,
social dimensions, and
regional perspectives.

For example, studies of information and telecommunications needs from the orientation of users can complement studies of the demand for telecommunications from the orientation of service providers. Studies of information use and utilization can identify a range of social and cultural factors which are likely to encourage or hinder the use of information provided by telecommunications and offer a sound basis for the development of products and services. Information and communication policy studies can be directed at different community, regional and government levels to facilitate the development of legislative frameworks designed to ensure that equitable benefits flow from advances in telecommunications.

The potential for inter- and transdisciplinary research in information science and telecommunications, and indeed other disciplines in the social sciences, stems from a number of issues which are germane to information science and which are relevant to a consideration of telecommunications as a key resource in the transmission of information as well as in the provision of access to information. Key issues are the nature of information, the nature of information behaviour, access to information products and services, and the evaluation of information products and services. The prospect of global telecommunications with all communication channels and media integrated into the one single digital medium has significant implications and far-reaching consequences for our understandings of human communication and human intelligence. Telecommunications research needs to begin addressing some of these implications and consequences. These four key issues are offered as part of this larger project.

1. The nature of information

Any discussion of information and communication technologies developed for the transmission of information begs the question: what is information? Although this has proven to be a challenging question for information scientists and some have advocated abandoning the search for an answer, there are three different but complementary descriptions which are relevant to this discussion of telecommunications research. Implicit in these definitions is the idea that information is used by people for a range of purposes: that is, information is a means to an end. All three descriptions reinforce the need for researchers to develop a broad understanding of information. Such an understanding is critical for any appreciation of the impact of the uneven spread of information technology and telecommunications, and consequently their benefits, not only among nations but also within nations.

Dervin and Nilan (1986) contrast two views of information. The objective view holds that information exists independently of those who know it and use it. Information is assumed to have constant meanings and correspondence to reality. Information can be compared to any other object in the sense that it can be located or placed in the physical world. By contrast, subjective information has no existence independently of those who know it. Information is essentially constructed by individuals and is the sense that people make of their worlds. Each of these views of information is at the core of the so-called traditional and alternative paradigms in information science. Studies of information in each paradigm assume different views not only of information, but also of information users and information use. In the traditional paradigm, which assumes the objective view of information, people are regarded as input-output information processors. Studies of information use explore people's interactions with formal information sources and systems. In the alternative paradigm, based on the subjective view of information, people are seen as active in seeking and using information for sense-making. Studies of information use in this paradigm explore the situations in which people need information as well as what lead to and follows their interactions with formal and informal information sources and systems.

In identifying three different uses of the term "information", Buckland (1990) relies on accepted definitions of the term as recorded in the Oxford English Dictionary. He suggests that the term is used to signify:

1. information-as-process which denotes the changing of what a person knows by being informed.
2. information-as-knowledge which denotes that which is changed in information-as-process.
3. information-as-thing which denotes objects such as data and documents which are regarded as being informative.

Buckland categorizes information as tangible, similar to Dervin and Nilan's view of information as objective, and as intangible, rather like the subjective view of information. He distinguishes also between entities and processes of information. He regards information-as-knowledge and information-as- thing as entities and information-as-process and information, or data, processing as processes.

In his description of information, Menou (1995:481) extends the ideas of Dervin and Nilan, and Buckland by addressing information, knowledge and wisdom. He categorizes information according to its various states, identified as:

- a product, which encompasses information-as-thing, as-object, as-resource and as-commodity;
- what is carried in a channel, including by extension the channel itself, as for instance assumed in expressions such as "electronic information", or "the media"; and
- the contents.

The contents state has three levels: semantic, syntactic and paradigmatic. It is the paradigmatic level which is most important when considering questions of the impact of information and telecommunications. It is this level which involves changes in a person's knowledge state through the making of meaning. These changes have obvious links with learning. Unlike most information scientists, Menou considers wisdom as part of the information-knowledge continuum and suggests that it consists of particular knowledge structures which can be individual and collective.

Apart from establishing the highly complex nature of information, these three descriptions reinforce the need to clarify and make explicit in research projects, if not in the research agenda itself, descriptions or definitions of information and assumptions about people who use and do not use information. Telecommunications transmit information-as-thing and increasingly information-as-commodity. Because people have the capacity to transform information into knowledge, information-as-knowledge and information-as-contents are a vital and legitimate research focus in telecommunications. Certainly studies of economic, policy and technological aspects of telecommunications need to continue but they need to be complemented and shaped by studies which place people and communities at the centre of telecommunications systems and services.

2. The nature of information behaviour

The adoption of a broad view of information assumes that human beings are active in their information behaviour: that is, they know when they need information and they are able to meet that need by using information. Information behaviour is contextually based and is strongly influenced by people's social roles as well as by their perceptions of their situations and problems. It is therefore crucial that information behaviour is conceptualized from the perspective of information users themselves rather than from the perspective of information and telecommunications providers and suppliers.

In the field of information science there are three particular models of information behaviour which can provide useful frameworks for telecommunications research. The first model, proposed by Wilson (1981), is based on a theory of motivation for information seeking and use. It suggests that physiological, affective and cognitive needs are at the core of information seeking behaviour and are shaped by the roles that a person fills in social life. For example, a person's work-role is carried out in the context of a work environment, a socio-cultural environment, a politico-economic environment and a physical environment. The second model, called SITUATION-GAP-USE, is based on a sense making approach to information needs and use. It suggests that information needs arise when an individual's sense has "run out" (Dervin and Nilan, 1986). The sense maker, or information user, is prevented from moving forward in a situation and faces a gap which must be bridged so that movement can be resumed. The model has been used to determine how people make sense of their worlds and how information helps them to do so. The third model, a model of information use environments developed by Taylor (1991:218), is informed by the choices people make about the information that is useful to them at particular times. The information use environment, or IUE, is a set of elements that "affect the flow and use of information messages into, within and out of any definable entity [for example an organization, an information system, a social group, a profession] and determine the criteria by which the value of information messages will be judged." The components of an IUE consist of sets of people, problems and their dimensions, settings and resolutions of problems which are dependent on information uses and information traits. In his description of the model Taylor acknowledges that it is informed in part by Dervin's observation that "systems personnel and researchers have been looking at something that they call information rather than at something users call information."

Studies of telecommunications demand can be complemented by studies of information needs, seeking and use. The essential difference between demand and needs studies seems to be the starting point, and whether it is an analysis of existing telecommunications products and services as a basis for the development of strategies to increase market share or whether it is an analysis of users' needs as a basis for the development of products and services which do more than merely repackage existing information. There are persuasive arguments for an approach which begins with the analysis of user needs and which takes account of the personal, social, educational and cultural factors embedded in the settings and contexts in which people use information and telecommunications.

An understanding of information seeking behaviour can also advance studies of the social dimensions of telecommunications. Much of the richness of human communication stems from the unplanned, non-directed, fortuitous interactions that occur. Creativity too is fostered by serendipity. Organizational development owes as much to half-formed ideas, tacit knowledge and intuition as it does to rigorous and systematic planning. Yet because these interactions and ways of knowing cannot be captured and stored in digital form there is a danger that they will be undervalued and ignored. Compared to the human brain, telecommunications offer a poor alternative to the processing of information. Current technologies require information to be organized in a linear sequential fashion and in a raw, de-humanized state for the purposes of transmission. Information transmitted through telecommunications is accessible in a limited number of ways and these are determined by telecommunications providers rather than by the users and potential users of telecommunication services. Telecommunication services are available in a limited number of locations and social settings, not all of which are necessarily conducive to reflective thinking or the exchange of creative ideas. As more and more information becomes available in digital form and as our interactions with others are carried out increasingly through electronic networks, areas for research will concern the impact of mediated experience on people's perceptions of reality and their sense of well-being.

3. Access to information products and services

A useful approach to issues of access is to consider a framework which allows the needs of telecommunications providers and users to be addressed in a coherent manner. Such a framework can also be used to address a range of policy, technological, social and educational matters as they relate to individuals, communities and government. One framework which has been applied to information services identifies different levels of access (Baker and Lancaster, 1991) and it has been modified here to suit a telecommunications environment. A constant factor across all the levels of access is social justice and equity, particularly for relatively isolated communities in metropolitan, urban and rural areas and low income-low traffic communities, especially in the developing world. Access issues in relation to telecommunications occurs at four levels.

1. Societal accessibility refers to the value that a society places on access to telecommunications especially for maintaining and developing social organization and cultural identity. Relevant questions for research might be: What policies, plans and strategies are in place at national and regional levels to ensure equitable access to telecommunications? What are the priorities for the development of telecommunications infrastructure and services? How does government discharge its responsibilities to its citizens in this regard? What, if any, are appropriate regulatory mechanisms for telecommunications providers and for content and infrastructure suppliers? How can the "greater good" be served in a market economy for telecommunications? How can competition among media and telecommunications conglomerates be harnessed to ensuring benefits to all? How can a balance between the global and the local be fostered and maintained? What is the role of telecommunications service providers in the delivery of information services? On what basis should technical standards be developed?
2. Institutional accessibility refers to social institutions and organizations which provide access to telecommunications for their members or clients. Relevant questions for research might be: How are social institutions such as museums, libraries and schools meeting their obligations in regard to telecommunications? How are the telecommunications carriers and content suppliers meeting their obligations? What support do cultural institutions require in order to contribute to the development of local and regional identity given the spread of telecommunications from the West? What should be the criteria for judgements on content carried by telecommunications? What roles do social institutions have in monitoring developments in telecommunications and in ensuring that those developments address social needs?
3. Intellectual and psychological accessibility refers to the "friendliness" and approachability of telecommunications. Relevant questions for

research might be: How culturally and gender inclusive are telecommunications products and services? How is the quality of telecommunications products and services assured? What are the assumptions made by product and service designers and developers about the interests, skills and capacities of people who use telecommunications? What design features enhance people's understanding of information delivered by telecommunications? How do telecommunications services reflect the cognitive and affective needs of the users of these services? How do telecommunications products and services address the real needs of people in innovative ways? Are education and training programs accessible to all users and potential users of telecommunications?

4. Physical accessibility refers to the ability of individuals and groups to access telecommunications services. Relevant questions for research might be: What are the criteria for providing physical access? How are the criteria derived and who sets them? What are desirable levels of access? How can the needs of groups with special needs, for example the elderly, the visually impaired, the physically impaired, be met? What are the roles of public libraries, community centres and schools in providing access? To what extent are occupational health and safety standards observed in the provision of telecommunications infrastructure and services? What are the conditions for interactivity in telecommunications and how are the conditions being met?

This framework for the consideration of issues of access takes considerable account of the social and cultural factors that were identified at the Workshop. It does so in such a way that these factors can be integrated into the development of telecommunications policy and its implementation at government, institutional and community levels. If people are the focus of future developments then there will be significant benefits to all. If the technologies are the focus, there is a real risk that the benefits will be restricted to a few and harmful to many. Ideally, developments should be driven by the needs of people as they themselves identify them. At the same time, it must be recognized that advances in technology occur in the context of the social and political world. The challenge for telecommunications research is to investigate and identify the minimum conditions under which telecommunications will provide widespread social and economic benefits.

4. Evaluation of information products and services

The discussion of the nature of information and information behaviour suggests that the evaluation of telecommunications should be two-fold. Firstly, evaluation of the infrastructure itself will need to involve community representatives, technical experts, and economic analysts in the assessment of the social usefulness of delivery mechanisms and the provision of social justice and equity. Judgements will also need to be made about the appropriateness of any policy frameworks as well as the effectiveness of the technologies. Secondly, evaluation of services delivered by the infrastructure will need to take place and it is this aspect of the evaluation which can be most appropriately informed by information science. The framework used for addressing issues of access provides some pointers for evaluation measures and suggests that there are two aspects of information products and services that need to be evaluated; the first is the information itself and the second is the impact of information on people.

The evaluation of information transmitted by telecommunications, or objective information, relies on external characteristics such as the authority and credibility of authors and publishers, the validity and reliability of the information provided and the quality of its presentation. Considerable progress has been made in the development of user-oriented measures for information services concerned with the delivery of information at organizational and institutional levels and telecommunications research could build on this work adapting it to the requirements of a digital information environment.

The evaluation of subjective information or information-as-knowledge relies on measuring the impact of information on people and their achievement of goals, completion of tasks or their sense making. It is possible to evaluate the impact of information on individuals, communities, regions or nations, with the major challenges being the development of conceptual frameworks and methodologies. Once these are established then appropriate indicators and measures can be developed and validated. Two essential research areas are the nature of the balance between the benefits and costs of telecommunications and the assessment of the extent of any positive differences made by telecommunications to people's lives.

The evaluation of the impact of subjective information is a longer-term process than the evaluation of objective information. Information-as-process may also require evaluation in which case indicators of information competence will need to be developed. Reporting on a project concerned with the impact of information on development, Menou (1993) outlined a set of principles for evaluation. They bear some similarity to the suggestions already outlined. Menou's principles include a prescription for beneficiary- or user-driven assessment of impact, the selection of indicators which will be used in any given situation, and a recognition that assessment of impact is an ongoing process. These principles are a helpful guide to the evaluation of the impact of information but also the impact of telecommunications.

This discussion of key issues in information science demonstrates that there are a number of significant, complementary and common areas for research in both information science and telecommunications. Engagement in cross- and trans-disciplinary research in these areas has the potential to enrich both disciplines. It now remains for researchers and funding agencies in telecommunications and information science to work towards realizing the full potential of those areas in the interests of the continuing development of human society.

The author wishes to acknowledge the valuable contribution made by Michel Menou in his comments on an earlier version of this paper.

References

Buckland, M.K., 1990, Information as thing, *Journal of the American Society for Information Science* 42(5):351-360.

Dervin, B. and M. Nilan, 1986, Information needs and uses, *Annual Review of Information Science and Technology* 21:3-33.

Baker, S. and W.F. Lancaster, 1991, *The Measurement and Evaluation of Library Services* 2nd ed. A(Information Resources Press, rlington, VA).

Menou, M.J. (ed.), 1993, *Measuring the Impact of Information on Development* (International Development Research Centre, Ottawa, Canada).

Menou, M.J., 1995a, The impact of information – I. Toward a research agenda for its definition and measurement. *Information Processing and Management* 31(4):455-477.

Menou, M.J., 1995b, The impact of information – II. Concepts of information and its value, *Information Processing and Management* 31(4):479-490.

Shera, J.H. and D.B. Cleveland, 1977, History and foundations of information science, *Annual Review of Information Science and Technology* 12:249-275.

Taylor, R.S., 1991, Information use environments, in: B. Dervin and M.J. Voight, eds, *Progress in Communication Sciences*, 10 (Ablex, Norwood, NJ) 217-255.

Wilson, T.D., 1981, On user studies and information needs, *Journal of Documentation* 37(1):3-15.

CHAPTER 13

Telecommunications and management: the knowledge dimension

Bill Martin

Introduction

The impact of telecommunications upon organizations and upon the work of those who manage them, has tended to have less exposure in the literature than has that of computers. This in itself is not important particularly in an age where the convergence of these once separate technologies is manifest strikingly by the advent of multimedia communications, in which a variety of message formats, print, voice and image can be transmitted simultaneously and in real time on a global basis. What is important is appreciation of the wider and growing significance of telecommunications technologies to all organizations, both as enablers and as shapers of form and process. As awareness of this significance has grown there has been a consequent shift of emphasis in management theory away from immediate impacts and outcomes and towards a search for understanding of the wider meanings and broader implications of these key technological developments. This stems not from any diminution in the relevance or significance of telecommunications technology within organizational environments, but rather its widespread acceptance as infrastructure and platform in all manner of enterprises public and private, profit or non-profit, large or small. In particular, it reflects both the intelligent nature of networks and the range and flexibility of customized services now available from competing carriers and a wide range of value-added providers, and the potential synergies that can arise between this intelligent technology and the emergence of knowledge-based organizations.

Traditional views of technology impacts

During the 1980s, McFarlan and McKenney[1] drew attention to the need to connect the hitherto separate *islands of technology* that existed in many organizations, calling for the use of networks to overcome dispersal and dilution of the benefits of the considerable investments involved. Clemons and McFarlan[2] likewise warned of the dangers of failing to respond to the threats and opportunities inherent in the new telecommunications technologies, with the simple message of the need to either *hook up or lose out*. These and a host of succeeding writers told and retold the stories of such companies as Baxter Healthcare (formerly American Hospital Supply) and American Airlines, whose use of online information systems was to prove so profitable and to spawn so many imitators. The examples can be multiplied many times over but it is worth emphasizing that none of them would have succeeded without the power of telecommunications networking. Today, in a world of global markets and continuing business turbulence, this lesson appears to have been well learned, with the MIT *Management in the 1990s* programme being perhaps the most comprehensive attempt to address the wider relationships between organizational change and developments in the information and communications technologies.[3] More recently the connections between the two have been examined by Bradley, Hausman and Nolan[4] and the message is coming across with even more force. Telecommunications continues to be a key enabling technology, a source of value-added both to other activities and in its own right.

The problem with all success stories is of course that many of them are highly context-specific and the product of a wide range of factors, of which technology is likely to be only one. As Gillespie and Williams pointed out, network innovations cannot be divorced from the organizations within which they are embedded, and their effect, therefore, is not uniform being subject to the influence of particular production, distribution, decision-making and coordination structures.[5]

Nevertheless nearly a decade ago, Gillespie and Williams could identify a shift in the role played by telecommunications in economic development, from one that was essentially *passive* to one that was *active*, as expressed in the areas of product, distribution, process, transactional activity and managerial activity.[6] The recurrent theme in all these activities is that of the radical and pervasive influence of telecommunications on the production of goods and services, by digitization of both the process and the outcome, upon the structure of both markets and organizations through its real time, global delivery mechanisms and its transforming impact on both time and distance and on managerial decision making.

Nowhere is the impact of digital telecommunications upon organizations better illustrated than in its contribution to global trade in both goods and services. Such activity is particularly intensive in the area of financial management, with production, marketing and distribution, design and research and development all heavily dependent upon advanced high-speed communications. Manufacturing firms can use corporate data flows to schedule production of product components in different parts of the world, and to vertically integrate their assembly and delivery to global markets. Where corporate data flows are exchanged between branches of the same firm, they facilitate existing arrangements and allow for efficiencies in communication. Where they take place between different firms this can involve data that embody new kinds of goods and services such as software, information storage and retrieval and data processing.[7] What is occurring involves a new kind of fusion between computing and telecommunications that will affect all companies, whether or not they have been significant users of technology in the past. Indeed, it is argued companies will have to make substantial investments in technology not for sustainable competitive advantage but simply to avoid strategic jeopardy.[8]

Networking technologies

The key to such developments lies in the integration both of carrier and of content in a world where telephone networks have long since ceased to be simply the domain of voice telephony and have become computerized information networks with in-built intelligence in the form of extremely sophisticated software. In an age of intelligent networks, the focus has shifted from the provision of terminals, lines and network facilities to providing total end to end solutions to customer problems, to issues of corporate information flow and its management.[9] These networks have been developed by adding intelligent databases to existing public network signalling architectures. The databases and new digital switches are placed on top of the existing network to provide new interfaces for customers and carriers. The resulting Virtual Network allows the development of new services and provides cost effectiveness and security and allows customers to control the distribution of information according to instructions placed within the network.[10] Over these networks — telephone, cable, cellular and satellite — flows the information, an increasing array of basic services, such as electronic mail and interactive video, and an equally wide range of applications, for example distance learning and teleworking, which provide the user with dedicated solutions to communications problems. Key technologies include:

* Integrated Services Digital Network (ISDN)
* Asymmetric Digital Subscriber Line (ADSL)
* Asynchronous Transfer Mode (ATM)
* Group Special Mobile (GSM)

ISDN is currently a major carrier of business traffic and is particularly suited to the needs of small and medium sized enterprises (SMEs), supporting the transmission of voice, data and even moving image over copper pairs. ADSL uses the latest digital compression techniques to provide distributive and interactive narrow-band services such as VHS-quality video over copper pairs to the home. ATM is being hailed as the new standard for broadband, multimedia services, with the capability to deliver vast quantities of voice, data and video at rates of up to 2.4 gigabits per second. Mobile continues to be a high potential vehicle for a growing range of value-added data communications services. However, it is the ability of these devices and services to combine into networks that makes them such a compelling force in the business world. This force takes effect in two specific ways, as the so-called *wired company* and as organizations whose operational structure is in the form of a network.

The wired company

Wired companies use networks of computers and telecommunications to connect people to people and people to data. They employ tools like electronic mail, teleconferencing and groupware to allow people to work together despite distance and time and almost without regard for departmental and even corporate boundaries. In wired companies, business can be conducted on-line and in real time, with information flows bypassing traditional organizational hierarchies and passing directly between those who need it. This not only cuts whole layers off traditional hierarchies but carries fundamental implications for management and control. In networked companies there is said to be less supervision of the content of overall performance. Leadership becomes more important than management, communication more urgent than command.[11]

The major benefits of wired companies are:

* Saving of time and money: through real time interactive operations
* Removal of barriers between departments
* Access to a wider pool of knowledge and expertise through information sharing
* Fewer filters in place between information and those who use it.[12]

The networked organization

The impact of information and communications technologies (ICTs) on organizations is still the subject of considerable debate. Rockart and Short identified a variety of impacts enabling the management of interdependence within and between organizations, and facilitating a range of organizational innovations up to and including Business Process Reengineering (BPR).[13] There is by no means unanimity among management writers on either the direction or the outcome of these interactions between technology and organizations. Some authorities would anticipate a greater rather than a reduced centralization of managerial power, while others would argue that flatter, more-responsive, team-based organizational structures are the model for the future. In view of the shortage of empirical evidence and the relative recency of the organizational innovations involved, it is too early to be categorical about the nature and potential longevity of these developments. Nevertheless, certain trends resulting from the impact of new technology on organizations are already apparent and, in addition to the emergence of networked structures, these would include the development of flexible working arrangements through teamwork, the increased integration of systems, processes and functions, and an attempt to leverage information and knowledge within the context of what are increasingly being regarded as learning organizations.

Teams and networks

There can be confusion as to the relationship between networks and teams, with some writers using the terms interchangeably and others distinguishing between the two. In some companies, *networks* imply a set of external relationships, a global web of alliances and joint ventures. In others, networks are seen as new ways for executives to share information through the use of groupware and similar tools. Another perspective is that of informal ties among managers, who work in floating *teams* that operate across functions and cut through the everyday rigidities of organizational hierarchy. In this form, one finds it hard to distinguish between *networks* and *teams,* the latter being defined as groups of people drawn from across the company's functions, business units and from different levels of the hierarchy in order to solve specific business problems. The essential difference seems to lie in longevity, in that whereas teams tend to be disbanded once the job is done, networks continue in existence indefinitely.[14]

Irrespective of differences in nomenclature however, the central characteristics of these organizational innovations are their effects on decision-making and on operational effectiveness. Networks are designed to empower people, and to enrich the quality of their decisions in a sharing, client-oriented environment that looks beyond immediate issues of narrow functional or departmental interest. At one level, this can involve the use of communications technologies to achieve integration across the firm's value-added chain, with implications not just for products, services and customer relationships, but also for the alignment of business and technology strategies within the organization. At another level, however, the decision to adopt a networked structure goes right to the heart of the organization, to its goals, mission and purpose. This involves the creation of a new *social architecture* for the organization, to support its development as a dynamic, information-based, people-centred organism.[15]

Information, the invisible resource

It is clear that as the 21st century draws near, management has learned the lesson that in an information-based economy, organizations must seek to be competitive by adding value to their two greatest assets, their information and their people. The emergence of new network-based organizations by no means signifies the end of formal structures for coordination and control. Rather, it is posited on recognition of the need to redefine such systems in order that they will support the organization in its attempts to leverage both its information and its people.[16]

Although information has always been a resource for management, traditionally it has been used in the context of bureaucracies, provided on a need-to-know basis through management information systems, and treated as something which was scarce and, therefore, which needed to be hoarded or kept secret. In the information age, however, not only is information recognized as a vital organizational resource, but also so is the fact that it has to be organized and disseminated at the corporate level if it is to be leveraged as organizational knowledge. With this realization has also come recognition of the fact that in many cases, the information available to managers is inadequate, whether in terms of format, timeliness or relevance. As a result, networking and other information technologies are now being employed in an attempt to generate, organize, communicate and share the information that organizations need if they are to be effective, competitive, profitable and relevant to the needs of their customers.[17]

Networks and people

If human networks are to be effective, those communications networks which they use to support the exchange of information and ideas must be sufficiently transparent as to enable all members of the network to have instant access to the same information. This addresses the widespread problem of incomplete and inappropriate information, and obviates the need for time wasting disagreements over the content or substance of messages.[18] The provision of access to corporate information has the effect of *empowering* members of the network, and facilitates the goal of engaging well-informed and knowledgeable people in collaborative projects. Empowerment means recognizing and releasing into the organization, the power that people already have in their wealth of useful knowledge and internal motivation.[19] This collaborative activity produces certain synergies between people and between people and information. Moreover, by motivating people to *buy into* and to *own* particular projects, it promises to get the most out of them as valuable *knowledge assets* in their own right. It is people who are the ultimate resource in organizations, particularly in those information and knowledge intensive organizations of the information age. If empowerment is to work, it requires a commitment to information-sharing and to collaborative goal setting with ultimately, replacement of the hierarchy by a system of self-directing teams. Many of today's complex decisions in fact require input from a collection of people. In seeking to achieve synergies between what people know and the change processes necessary for organizations to survive and prosper in the information age, the links between people, information, knowledge and learning are being further extended.

Learning and the intelligent organization

The MIT *Management in the 1990s* research programme suggested that the *metabolic rate* of the organization, that is the rate at which information moves and decisions are made, is speeding up and will get faster in the 1990s.[20] A couple of years later, Haeckel and Nolan defined a company's *complexity* as a function of how many information sources it needs, how many business elements it must coordinate, and the number and type of relationships that exist among those elements. They then defined *corporate IQ* as a company's ability to deal with complexity, that is, its ability to capture, share and extract meaning from marketplace signals.[21]

Haeckel and Nolan advocated the creation of models of business enterprise that would represent the operations of an entire business, and which could then be captured in software programs which could be used to help pilot the business in much the same way as modern fly-by-wire aircraft are flown. These models would have an in-built capacity to change based on signals from the environment, in other words to exhibit elements of learning behaviour.[22] Haeckel and Nolan defined *institutional learning* as the process by which information models change. McGee and Prusak offer a more detailed description of the process, stating that organizations learn in two ways. First, organizations learn from the ways they support or promote learning by their individual members. Second, organizations learn from the ways they create systems and relationships to leverage individual learning to meet broader organizational goals.[23]

It is clear that telecommunications platforms can be a key element in both individual and organizational learning, whether through remote access to tools and sources or by means of interactive instructional packages in the home or office. It is also the case, that the intelligent networks themselves can play a part, both by adding to the sum of individual and corporate knowledge and by assisting in leveraging that knowledge in pursuit of organizational aims and objectives. These networks are also marketplaces in their own right involving the sale of value-added and data services, remote access to databases, banking and financial services, education and entertainment. In Australia, recent investments in cable networks by both Telstra and Optus, and their respective pay television alliances with global players in the information and entertainment industries underline the social and commercial significance of telecommunications. However, nowhere is this more readily apparent than in those developments in global networking taking place on the Internet.

Management and the Internet

Such is the pace of its growth, that it is clearly an idle, not to say futile exercise to speculate as to the size or extent of the Internet. Suffice to say that the Internet is a network of thousands of computers utilizing a common set of technical protocols to create a worldwide communications medium. Something of the order of 20 million people are using this network in over 70 countries around the globe, and the growth rate is predicted at around 10 per cent of this total user base every month. These users range from individual enthusiasts to national governments, from multinational corporations to non-profit organizations, from research centres and academic institutions to sporting organizations and corner shops.

The Internet is an outgrowth of a project launched in the 1970s by the United States Department of Defense's Advanced Research Projects Agency, the ARPANET. From this project emerged the TCP/IP suite of communications protocols and these were followed by basic services for remote connectivity, file transfer and electronic mail. In 1990, the ARPANET was taken over by the NSFNET and funded by the National Science Foundation. Although originally a research and educational network, these institutions now account for only one third of Internet traffic, the remainder being accounted for by commercial organizations.

In Australia, access to the Internet is provided by AARNet, the Australian Academic and Research Network, a private data network run on behalf of the Australian Vice Chancellors' Committee on lines leased from Telstra. On account of the sheer pace of growth in network traffic, which is doubling every seven to eight months, management of the AARNet has passed to Telstra. Educational and research institutions make use of the Internet for access to information, for file transfer and for communication by electronic mail. Commercial interest in the Internet has been stimulated, as elsewhere, by the potential for closer links with business partners, vendors and information resources via the network. Specific advantages for business on the Net include:

* Opportunities for the creation of a client base
* Product analysis
* Market analysis
* Rapid communications
* New business opportunities
* Cost-effective document transfer
* Human resource management
* Project management
* Collaborative work groups
* Product distribution
* Linkages with customers and vendors

Perhaps the major impetus behind the growth of electronic commerce in Australia in recent years has been availability of the World Wide Web, a hypertext tool which provides access to multimedia information and to complex documents and databases on the Net. Commercial interest has also been sparked off by recent decisions to liberalize access to the Internet and in particular, to allow commercial firms to set up in business as access providers.

In any event, there are clear signs that electronic commerce is developing steadily in Australia. This would include the emergence of completely new companies, formed to conduct business on the network. It would include the extension of existing business activities to the network marketplace, and it would reflect the convergence both of carrier and content, and of once separate communications media as innovative examples of electronic commerce. Current activity on the World Wide Web by Australian companies can be categorized as follows: Internet access; network consultancy; electronic publishing/interactive multimedia; advertising and promotion via home pages; the sale of goods and services via electronic shopping malls; and the conduct of everyday business in an electronic environment.

Whereas the advantages of Internet commerce are apparent, there are still obvious pitfalls to a major Australian presence on the electronic marketplace. In the first place, the cost of network access remains high, with Austel the telecommunications industry regulator being critical of the high cost of network services levied by both carriers, Telstra and Optus. This problem may well be addressed by the deregulation of telecommunications provision in 1997. Secondly, there are the limitations on Australian participation in the extremely lucrative market for information content, be this in online information provision or in interactive multimedia products and services. Immediate problems concern the prohibitive cost of producing material for the small

Australian market, and the fact that ownership of much of this content lies in foreign hands. The most realistic response to the latter problem seems to lie in strategic alliances with overseas content owners and the attraction of major international players to Australian locations.[24]

Conclusion

Telecommunications are critical to the successful operation of all kinds of organizations in this information age. They serve as enablers and as global platforms for design, research and development, manufacturing, marketing, and the distribution of goods and services. They facilitate organizational restructuring be it into networks of self-directed teams, or through the emergence of *virtual organizations* whose metier and *raison d'être* is the network. They hold the promise for organizations of being able to leverage their investments in technology, information and people in order to compete in an economy in which knowledge and organizational learning are key resources. For all these reasons, telecommunications is now a key element in business education around the world, and a pervasive topic in the management literature. The significance of telecommunications at the wider human and social level is likely to become more apparent as nations seek to exploit the Infobahn and its myriad business opportunities by transforming their educational systems, their industries and their commercial and communications infrastructures for participation in the global information economy of the 21st century.

As such it is likely to remain on the research agenda for the forseeable future.

Notes

1. There were three papers on the subject:
   F. Warren McFarlan and James L. McKenney, 1982, The information archipelago: maps and bridges, *Harvard Business Review*, September-October, 109-119.
   F. Warren McFarlan and James L. McKenney, 1983, The information archipelago: plotting a course, *Harvard Business Review*, January-February, 145-156.
   F. Warren McFarlan, James J. McKenney and Philip Pyburn, 1983, The information archipelago: governing the new world, *Harvard Business Review*, July-August, 91-99.
2. Eric Clemons and F. Warren McFarlan, 1986, Telecoms: hook up or lose out, *Harvard Business Review*, July-August, 91-97.
3. Michael S. Scott Morton, ed., 1991, *The Corporation the 1990s: Information Technology and Organisational Transformation* (O.U.P., New York).
4. Stephen P. Bradley, Jerry A. Hausman and Richard L. Nolan, 1993, *Globalization, Technology and Competition: the Fusion of Computers and Telecommunications in the 1990*s (Harvard U.P., Boston).

5. Andrew Gillespie and H. Williams, 1987, Telecommunications and the reconstruction of regional comparative advantage, *Environment and Planning* 20:1311-1321.
6. *Ibid.*
7. Sara Schoonmaker, 1993, Trading on-line: information flows in advanced capitalism, *The Information Society*, 9:39-49.
8. Bradley *et al.*, 1993 , *op.cit.*, 6.
9. Fred Symons, 1992, Prospects for technology, in: Anna Bodi, Anthony Newstead and Peter B. White, eds, *Telecommunications Management and Policy* (Monash Graduate School of Management, Melbourne) 147-154.
10. Ray Freeman, 1992, Intelligent networks and private betworks, in: Bodi *et al.*, *op.cit.*, 155-166.
11. Thomas A. Stewart, 1994, Managing in a wired company, *Fortune* 130(1):11, July, 16-22.
12. *Ibid.*
13. J.L. Rockart and J. Short, 1991, The networked organisation and the management of interdependence, n: Michael S, Scott Morton, *op.cit.*, 189-219.
14. Ram Charan, 1991, How networks reshape organisations — for results, *Harvard Business Review*, September-October, 104-115.
15. Christopher A. Bartlett and Sumantra Ghosal, 1995, Changing the role of top management: Beyond systems to people, *Harvard Business Review*, May-June, 132-142.
16. *Ibid.*
17. *Ibid.*
18. Charan, 1991, *op.cit.*, 112.
19. W. Alan Randolph, 1995, Navigating the journey to empowermentm *Organisational Dynamics* 23(4, Spring):19-31.
20. Scott Morton, 1991, *op.cit.*, 18.
21. Stephen H. Haeckel and Richard L. Nolan, 1993, Managing-by-wire, *Harvard Business Review*, September-October, 122-132.
22. *Ibid.*
23. James McGee and Laurence Prusak, 1993, *Managing Information Strategically* (John Wiley, New York) 208.
24. Roger Buckeridge and Terry Cutler, 1994, *Commerce in Content: Building Australia's Future in Interactive Multimedia Markets* (Cutler & Company, Melbourne) 44 pp.

CHAPTER 13

Comment: Donald Lamberton

As Martin notes, network innovations cannot be divorced from the organizations within which they are embedded, with the consequence that their effect is dependent upon specific circumstances and is unlikely to be uniform. This recognition of the importance of what goes on inside the organization bears on business strategy; and, in fact, upon the functioning of any organization.

Elsewhere (Lamberton, 1995) I have drawn attention to the prevailing practice in regulatory matters of treating both the firm and the regulator as black boxes, rather than complex organizations with informational requirements. Much more needs to be done to achieve a reasonable understanding of the effects of information asymmetry. Some concede that the task is unfinished (e.g. Baron) but there is a danger that having concluded that regulatory mechanisms, e.g. along Vogelsang-Finsinger lines, can be designed and judged that they can operate effectively, the problem of information asymmetry will be set aside.

This deserves emphasis because it has been "shown that slight perturbations in the standard information assumptions drastically changed all the major results of standard neoclassical theory. The theory was simply not robust at all" (Stiglitz, 1994:x).

References

Baron, D.P., 1989, Design of regulatory mechanisms and institutions, in: R. Schmalensee and R. Willig, eds, *Handbook of Industrial Organization*, Vol.2 (North-Holland, Amsterdam) 1347-1437.

Lamberton, D.M., 1995, Regulatory systems design, *Prometheus* 13(2):184-190.

Stiglitz, J.E., 1994, *Whither Socialism?* (MIT Press, Cambridge, MA).

Participants in the Workshop

James Alleman, University of Colorado at Boulder, US

David Allen, East-West Center, Hawaii, US

Cristiano Antonelli, Universita degli Studi di Torino, Italy

Colin Blackman, Editor, "Telecommunications Policy", UK

Martin Cave, Brunel University of West London, UK

Chris Chea, BTCE, Australia

Karen Clarke, National Library, Australia

David Cracknell, BT, UK

Mike Cronin, BTCE, Australia

Alain de Fontenay, TELA Group, US

John de Ridder, Telstra, Australia

Gunnar Eliasson, Industriens Utredningsinstitut, Sweden

Hans-Jürgen Engelbrecht, Massey University, New Zealand

Ove Granstrand, Chalmers University of Technology, Sweden

Maurice Haddad, BTCE, Australia

Peter Hagström, Harvard Business School, US

Nikola Holcer, World Bank, US

Marc Ivaldi, Institut d'Economique Industrielle, France

Richard Joseph, University of Wollongong, Australia

Meheroo Jussawalla, East-West Center, Hawaii, US

Joyce Kirk, University of Technology, Sydney

Donald Lamberton, Australian National University, Australia

Cliff Law, National Library, Australia

David Luck, BTCE, Australia

Gary Madden, Curtin University of Technology, Australia

Rodney Maddock, LaTrobe University, Australia

Bill Martin, Royal Melbourne Institute of Technology, Australia

Ian Martin, BTCE, Australia

Michel Menou, CIDEGI, France

Ann Moyal, Canberra, Australia

Karl-Heinz Neumann, Wissenschaftliches Institut für Kommunikationsdienst GmbH, Germany

John Nightingale, University of New England, Australia

Hajime Oniki, Osaka University, Japan

Almarin Phillips, Econsult, US

Francois Prothais, France Telecom, France

Deanna Campbell Robinson, University of Oregon, US

Mark Schankerman, London School of Economics, UK

A.J. Shaw, Department of Communications and the Arts, Australia

Marcellus Snow, University of Hawaii at Manoa, US

Kym Starr, BTCE, Australia

Diane Steinour, National Telecommunications and Information Administration, US

John Sutton, BTCE, Australia

G.M. Peter Swann, London Business School, UK

Lilian van der Vaart, Elsevier Science Publishers, The Netherlands

Sam Weerahandi, Bell Communication Research, US

Supplementary References

In view of the availability of Marcellus Snow's review of telecommunications literature and the extent to which it has been updated in the course of the various contributions to this book, it was decided to confine this listing to supplementary published references deemed by individual participants to have special relevance for the research frontiers theme.

Snow's review is available as:

Snow, Marcellus S., 1988, Telecommunications literature: a critical review of the economic, technological and public policy issues, *Telecommunications Policy* 12(2):153-183.

The following books are of considerable value for reference purposes:

Antonelli, Cristiano, ed., 1992, *The Economics of Information Networks* (Elsevier Science Publishers, Amsterdam).

Bohlin, Erik and Ove Granstrand, eds, 1994, *The Race to European Eminence: Who are the Coming Tele-Service Multinationals?* (Elsevier Science Publishers, Amsterdam).

Lamberton, Donald M., ed., 1995, *Beyond Competition: The Future of Telecommunications* (Elsevier Science Publishers, Amsterdam).

Lamberton, Donald M., ed., 1996, *The Economics of Communication and Information* (Edward Elgar, Cheltenham UK).

Mansell, Robin, 1993, *The New Telecommunications: A Political Economy* (Sage Publications, London).

Pogorel, Gérard, ed., 1994, *Global Telecommunications Strategies and Technological Changes* (Elsevier Science Publishers, Amsterdam).

Taylor, Lester D., 1994, *Telecommunications Demand in Theory and Practice* (Kluwer Academic Publishers, Dordrecht).

Attention is drawn also to the following journals as regular sources of research material:

Communications & Strategies

Information Economics and Policy

Prometheus

Telecommunications Policy

Supplementary References

Backhouse, R., 1982, *Information Services for Trade Unionists*, British Library R&D Report No. 5695 (Elm Publications, London).

Babe, R.W., ed., 1994, *Information and Communication in Economics* (Kluwer Academic Publishers, Dordrecht).

Baldasty, G.J., *The Commercialization of News in the Nineteenth Century* (University of Wisconsin Press, US).

Bud-Frierman, L., ed., 1994, *Information Acumen: The Understanding and Use of Knowledge in Modern Business* (Routledge, London).

Brynjolfsson, E., 1994, Information assets, technology, and organization, *Management Science* 40(12):1645-1662.

Computer Science and Telecommunications Board, National Research Council, 1994, *Realizing the Information Future: The Internet and Beyond* (National Academy Press, Washington DC, US).

Cronin, B. and N. Tudor-Silovic, eds, 1990, *The Knowledge Industries: Levers of Economic and Social Development in the 1990s* (Aslib, London).

Davies, A., 1994, *Telecommunications and Politics: The Decentralised Alternative* (Pinter Publishers, London).

Drake, W.J., ed., 1995, *The New Information Infrastructure: Strategies for U.S. Policy* (Twentieth Century Fund Press, New York).

Duch, R.M., 1991, *Privatizing the Economy: Communications Policy in Comparative Perspective* (University of Michigan Press, Ann Arbor).

Dudley, L.M., 1991, *The Word and the Sword: How Techniques of Information and Violence have Shaped our World* (Blackwell, Oxford).

Eliasson, G., S. Folster, T. Lindberg, T. Pousette and E. Taymaz, *The Knowledge Based Information Economy* (Industrial Institute for Economic and Social Research, Stockholm).

Fischer, C., 1992, *America Calling: A Social History of the Telephone to 1940* (University of California Press).

Freeman, C. and L. Soete, 1994, *Work for All or Mass Unemployment: Computerised Technical Change into the 21 Century* (Pinter Publishers, London).

Graham, S., J. Cornford and S. Marvin, 1996, The socio-economic benefits of a universal telephone network: A demand-side view of universal service, *Telecommunications Policy* 20(1):3-10.

Himmelstrand, U., ed., 1992, *Interfaces in Economic and Social Analysis* (Routledge, London).

Huberman, B.A. and T. Hogg, 1995, Distributed computation as an economic system, *Journal of Economic Perspectives* 9(1):141-152.

Ironmonger, D.S. 1972, *New Commodities and Consumer Behaviour* (Cambridge at the University Press).

Jussawalla, M., ed., 1993, *Global Telecommunications Policies: The Challenge of Change* (Greenwood Press, Connecticut, US).

Jussawalla, M., ed., 1995, *Telecommunications: A Bridge to the 21st Century* (Elsevier Science Publishers, Amsterdam).

Lamberton, D., 1994, The information revolution in the Asian-Pacific region, *Asian-Pacific Economic Literature* 8(2):31-57.

Macdonald, S., ed., 1996, *Special Issue on Informal Information Flow, International Journal of Technology Management* 11(1-2).

Machlup, F. and U. Mansfield, eds, 1983, *The Study of Information: Interdisciplinary Messages* (John Wiley, New York).

Maguire, C., E. Kazlauskas and A. Weir, 1994, *Information Services for Innovative Organizations* (Academic Press, San Diego, US).

Malone, T.W., J. Yates and R.I. Benjamin, 1987, Electronic markets and electronic hierarchies, *Communications of the ACM* 30(6):485-497.

Marvin, C., 1988, *When Old Technologies Were New: Thinking about Electric Communication in the Late Nineteenth Century* (Oxford University Press, Oxford).

Mansell, R., ed., *Management of Information and Communication Technologies: Emerging Patterns of Control* (Aslib, London).

Noam, E., S. Komatsuzaki and D.A. Conn, eds, 1994, *Telecommunications in the Pacific Basin: An Evolutionary Approach* (Oxford University Press, New York).

Perelman, M., 1991, *Information, Social Relations and the Economics of High Technology* (Macmillan, London).

Preston, L.E. and D. Windsor, *The Rules of the Game in the Global Economy: Policy Regimes for International Business* (Kluwer Academic Publishers, Norwell, US).

Sauer, C., 1993, *Why Information Systems Fail: A Case Study Approach* (Alfred Waller, Henley-on-Thames, UK).

Steinfeld, C., J. Bauer and L. Caby, eds, *Telecommunications in Transition — Policies, Services and Technologies in the European Community* (Sage Publications, Thousand Oaks, US).

Stiglitz, J.E., 1994, *Whither Socialism?* (MIT Press, Cambridge, MA, US).

Tehranian, M. (1990), *Technologies of Power: Information Machines and Democratic Prospects* (Ablex, Norwood, US).

Thomsen, E.F., 1992, *Prices and Knowledge: A Market-Process Perspective* (Routledge, London).

Thomson, R., ed., 1993, *Learning and Technological Change* (Macmillan, London).

Wasco, J., V. Mosco and M. Pendakur, eds, 1993, *Illuminating the Blind Spots: Essays Honoring Dallas W. Smythe* (Ablex, Norwood, US).

Index of Authors

Keyword Index